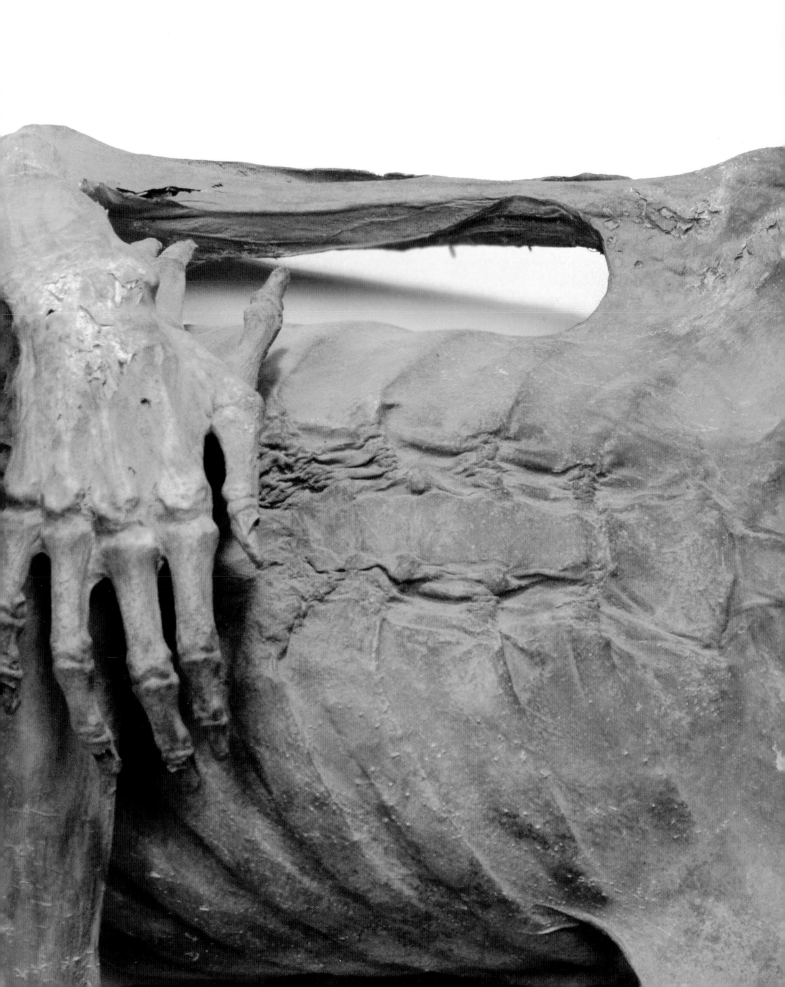

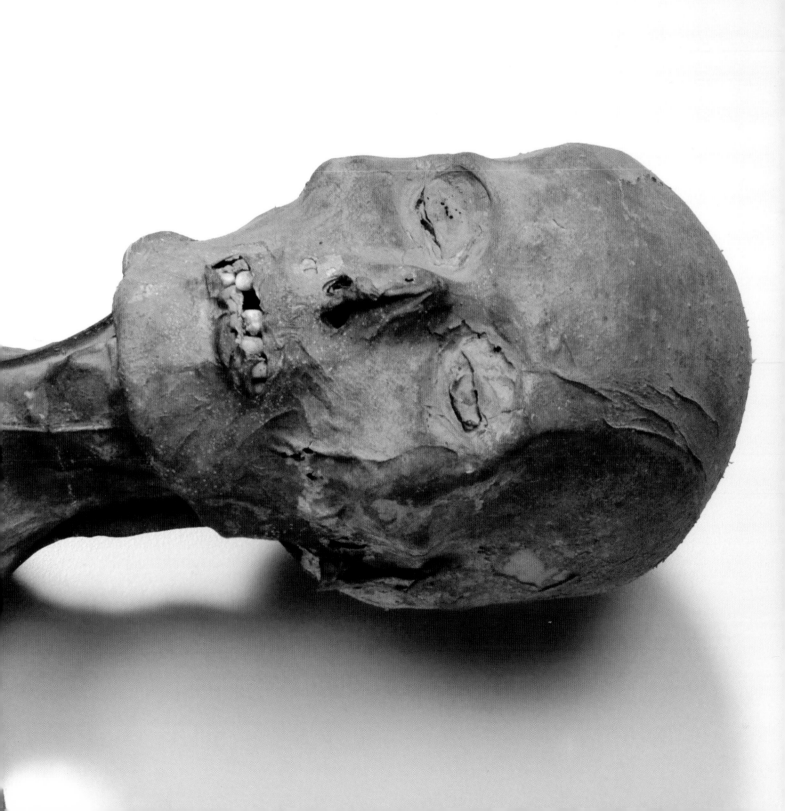

LONG LIVE THE DEAD:

THE ACCIDENTAL MUMMIES OF GUANAJUATO

Written by: Louis Aguilar
Afterword by: Kevin Prihod

First edition published by Accidental Mummies Touring Company LLC in October 2009 as the companion to *The Accidental Mummies of Guanajuato* exhibition produced by Detroit Design & Exhibits at Eekstein's Workshop LLC, a wholly owned subsidiary of the Detroit Science Center, in association with Accidental Mummies Touring Company LLC.

For information, address: Detroit Science Center
　　　　　　　　　　　　5020 John R Street
　　　　　　　　　　　　Detroit, MI 48202

ISBN: 0615316689

ACKNOWLEDGEMENTS:

Book Production:	Kelly Fulford, Detroit Science Center
	Christina Pattyn, Ciel Design Partners
Content Developer:	Martina Guzman
Forensic Captions:	Vivian Henoch
Design:	Michael Ulinski, Ciel Design Partners
Photography & Illustration:	Barbara Martin Bailey
	Christine Chambers
	Edward Gorczyk
	Janet Jarmon
	Vivian Henoch
	Julio Zenil
	(SEE CREDITS, PAGE 152)
Editor:	Greg Tasker
Printed by:	Colortech Graphics – Roseville, MI

LONG LIVE THE DEAD:

THE ACCIDENTAL MUMMIES OF GUANAJUATO

Louis Aguilar

Afterword by Kevin Prihod

The Official Companion to the Exhibition

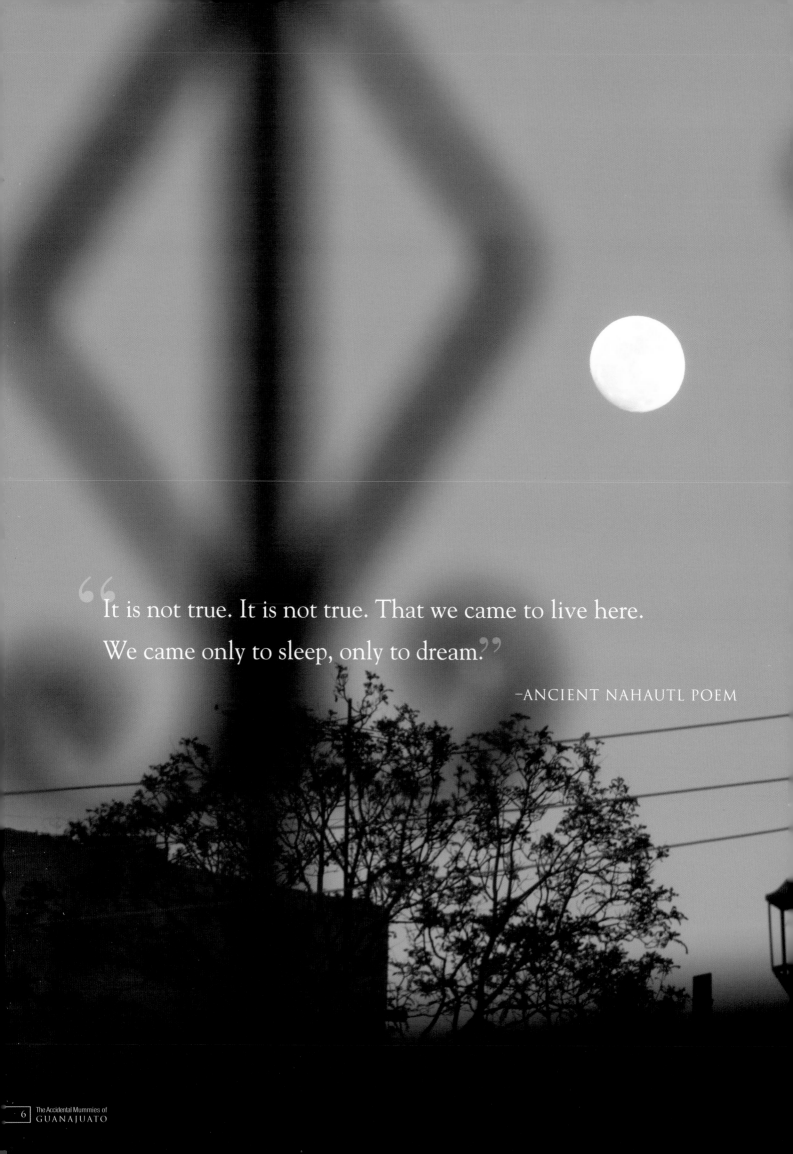

"It is not true. It is not true. That we came to live here.
We came only to sleep, only to dream."

−ANCIENT NAHAUTL POEM

introduction:
LONG LIVE THE DEAD:
GUANAJUATO'S ACCIDENTAL MUMMIES

One late afternoon in the Guanajuato cemetery where mummies are accidentally born, mariachis sang sad romantic songs to a headstone. Beside them, a hunched old woman cried and laughed and talked to the grave.

Mariachis often serenade the dead at *Panteón Municipal Santa Paula*, the city's crowded old graveyard. This is Mexico, after all, where indigenous rituals have tangled up with Catholic ceremony to create a society that revels in elaborate, bittersweet celebrations.

One is Day of the Dead, a November affair in which millions seek to communicate with loved ones who have passed away. Parades, music, food, joy and pathos fill the day. It's simply disrespectful to allow the dead to fade from the psyches of the living, many Mexicans believe.

But a phenomenon in Santa Paula challenges many people's beliefs about the dead: The Accidental Mummies.

Accidental because no one planned for them to be mummies, unlike the Egyptian pharaohs. Accidental because they were only discovered due to a grave tax that caused the removal of bodies if the fee was not paid.

It was an accident that occurred 112 times. The first mummy – a French doctor – was uprooted in 1865. The last – a 9-month-old boy – was unearthed in 1958. Some initially feared the mummies could be demons. But that has never stopped millions from wanting to see them.

First, the mummies were stored to rest in catacombs beneath the cemetery. That eventually evolved into El Museo de las Momias. In years past, some regarded the mummies as a freak show and many today still consider them too creepy to see. Nevertheless, they attract artists and academics, and marketers and new age spiritualists. Mainly, they attract the merely curious.

Always in the streets of Guanajuato, sugary confections and trinkets resembling mummies can be bought. Fake cardboard mummies topped with sombreros capture the attention of tourists looking for a Kodak moment. Masks of the mummies are staples of celebration – weddings, wakes and art openings.

The true identities of most of the mummies are lost with time. Legend and lore fill the gap. One mummy is said to be a witch. Another may have been buried alive – speculation inspired by the intensity of her alleged frozen scream and the way her outstretched arms cover her face.

No one denies the mummies are fierce holdouts to absolute decay. Every drop of liquid and working organ has long evaporated, but they are far more than skeletons. Their skin is the color of parched paper. Several women have long braided hair. Most men's genitalia are intact. Tattered pieces of clothing survive – a single stocking, a pair of black boots – while moths eat away abdomens.

Baby girls are dressed like angels, and baby boys like saints. El Museo de las Momias, which has expanded through the years, is a sleek, modernistic building with ambient, shadowless lighting and climate-controlled rooms. So now the resting place of the mummies is similar to a trendy boutique hotel.

Year after year, attendance rivals that of a Broadway hit.

WALK AMONG THE DEAD
El Museo de las Momias offers plenty of close-up views of Guanajuato's mummified citizens. About 500,000 people visit the museum annually.

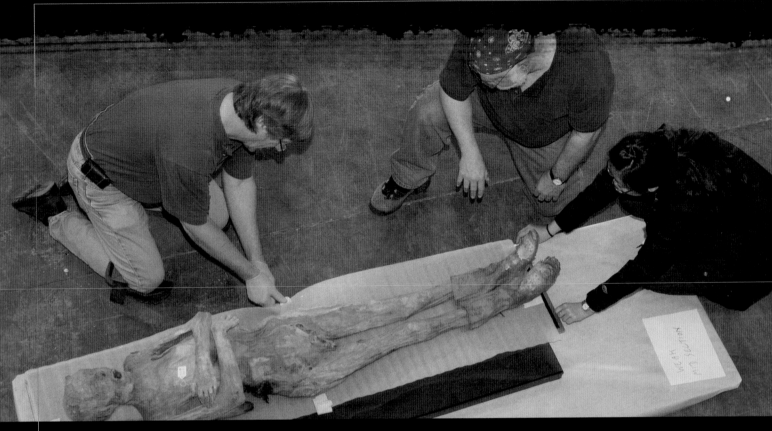

WHAT CAN THE MUMMIES TELL US?

Plenty. Former Mummy Road Show hosts and scientists Ron Beckett, on the far left, and Jerry Conlogue, along with research assistant Jiazi Li were invited by the Detroit Science Center to examine the mummies. The trio applied their knowledge at the Detroit Science Center's Design & Exhibits workshop.

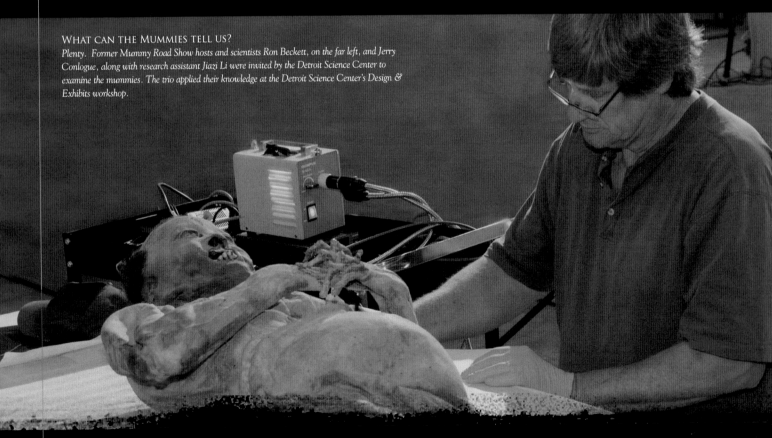

Possibly the last group to learn how to pay homage to the mummies are Western scientists, who were long regarded as the bogeyman in many communities.

For decades, anthropologists and archaeologists fleeced native cemeteries for skeletons and mummies and hauled them to universities and museums.

Enter Dr. Eduardo Romero Hicks, a U.S.-trained physician who became mayor of Guanajuato. In 2007, he invited three American scientists to apply science and advanced medical technology in the examination of the 22 mummies.

All are mummy devotees. Ronald Beckett and Jerry Conlogue, co-directors of the Bioanthropology Research Institute at Quinnipiac University in Hamden, Conn., once traveled North America in a former potato chip truck to examine any mummy any community would let them. They co-hosted something called *The Mummy Road Show* for the National Geographic channel. Beckett applied his deft skills of endoscopy: those slender tubes with tiny cameras at the end. Conlogue employed radiography: x-rays. The result is a far deeper look inside the mummies than previously attempted.

Jerry Melbye is a forensic anthropologist and a professor at the University of North Texas. He has studied mummies in Egypt's Sahara Desert, Mesa Verde in the American Southwest, and Vancouver Island, British Columbia. He took the information gained from the internal pictures, along with what he had learned from surface autopsies, to determine the basics of the mummies; approximate age, diet and cause of death.

In 2009, 36 mummies embarked on a North American tour thanks to the Detroit Science Center...

MUMMIES ON THE MOVE.
The Detroit Science Center's Design & Exhibits workshop became a hub of science and media attention months before the Accidental Mummies exhibit opened publicly. The 36 mummies were stored in climate-controlled rooms and handled with extreme care.

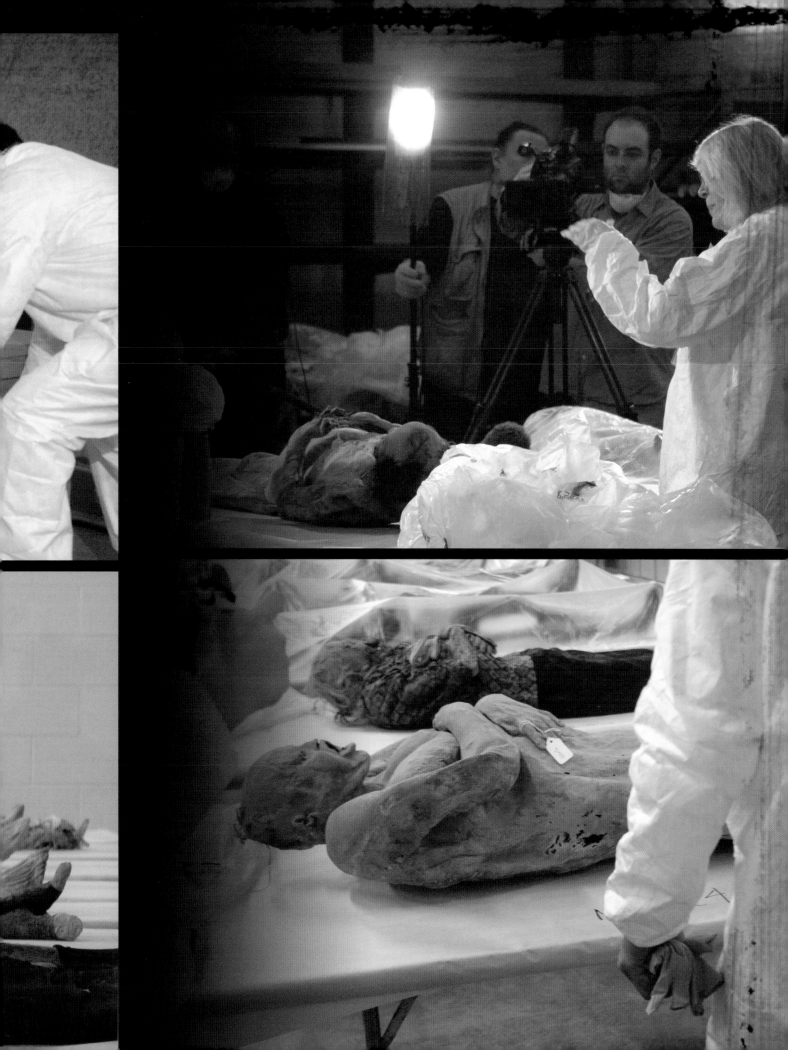

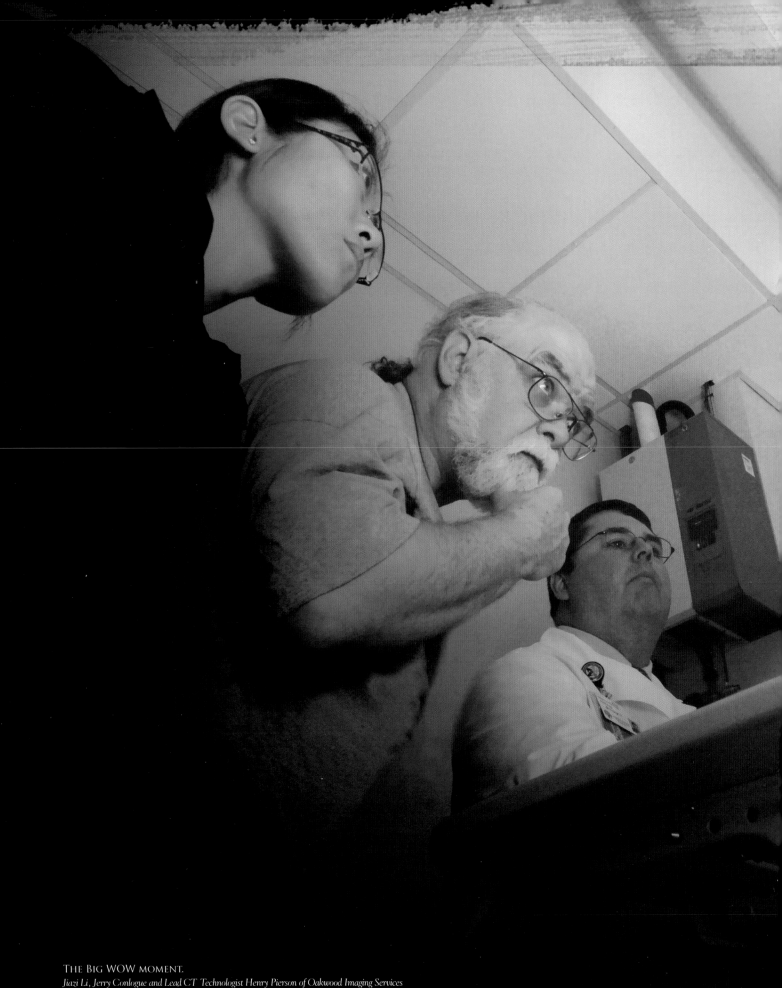

THE BIG WOW MOMENT.
Jiazi Li, Jerry Conlogue and Lead CT Technologist Henry Pierson of Oakwood Imaging Services marvel at the computer images generated by the full body CT scans of the mummy known as La Bruja (the Witch). Oakwood Hospital and Medical Center welcomed seven mummies into their imaging center in Dearborn, Mich. The precise images are often ethereal and stark.

In 2009, 36 mummies embarked on a North American tour thanks to the Detroit Science Center. The Detroit museum also is advancing what science can tell us about the mummies. The museum has reassembled the three members of the Guanajuato mummy team. And they have enlisted experts at Oakwood Healthcare System and the acumen of medical manufacturer Siemens Healthcare to examine the mummies via CT scan, commonly known as cat scans.

The CT scans captured tens of thousands of three-dimensional images that are a giant leap of data. Conlogue and Beckett were as giddy as school kids as they watched computer monitors unveil the images for the first time.

"My God isn't that beautiful?" Beckett said as he watched images from deep inside the ravages of a mummy known as the Witch.

They are beautiful. The hollowed insides of the mummies are as serene as celestial maps. But the deep lacerations and dislocations of the bones can look as volatile as a raging sea. The clarity of the images is so detailed they resemble anatomical sketches that hang in fine art museums.

The mummies have so many stories to tell us – not just about themselves, but about the way we treat the dead and how we live.

It's our job to understand how to ask. ❧

MARIANO HALLER
MURIO A LA EDA
DE 5 AÑOS 8 M
EL 13 DE SET
EUGENIA VILLANUE
FEBRERO 2
TORIBIO BAL
DOLORES DELGADO
ABRIL 24 D
PERPETU
MA. LUZ
DE
18-V-1897
RADA
ANA M
DA. DE

THE WITCH

La Bruja is placed against a wall of crypts in the Guanajuato cemetery that produced the accidental mummies. Little is truly known about the female mummy, but local lore says the woman was a good witch in real life.
The Accidental Mummies exhibit is the first time the Guanajuato mummies have been allowed to leave Mexico.

"To represent a masterpiece of human creative genius."

– CRITERIA TO BE SELECTED A WORLD HERITAGE SITE BY UNESCO.

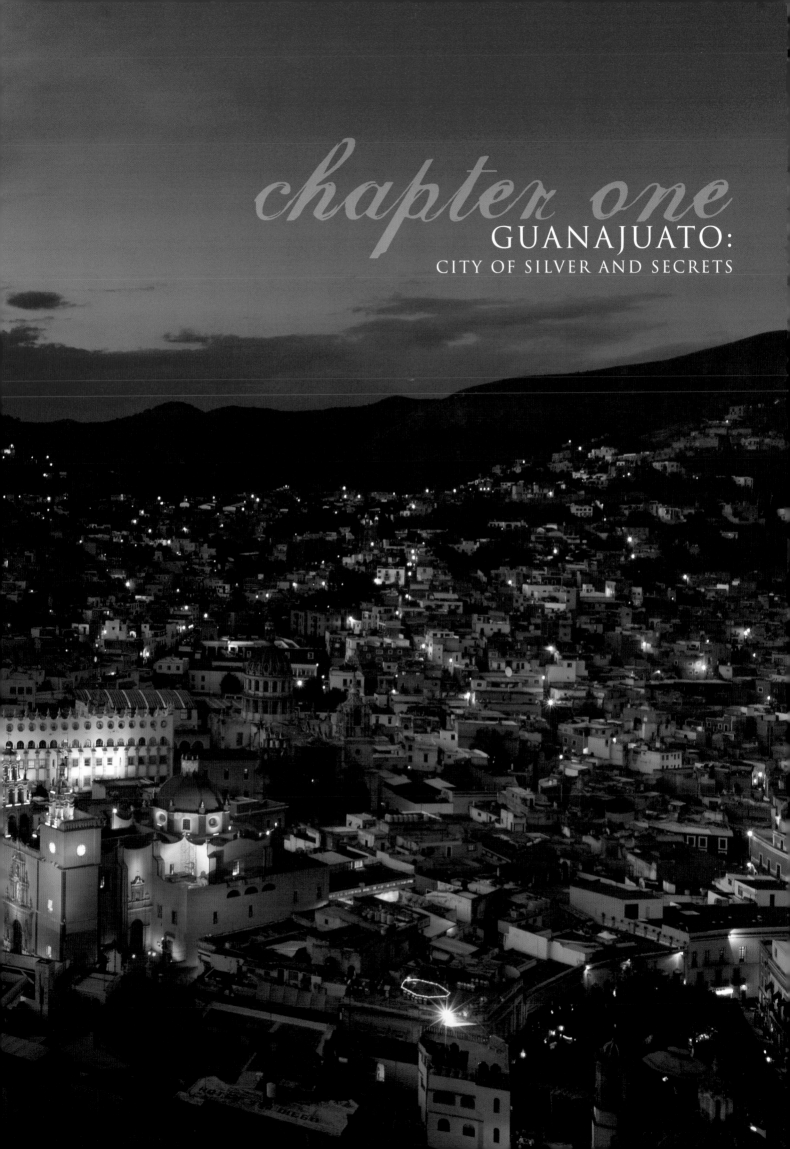

chapter one

GUANAJUATO:
CITY OF SILVER AND SECRETS

You enter the heart of Guanajuato by descending into long, underground tunnels of stone, which are the first signs this mid-size city has made a curious pact with its geography. The dim lights and the dust kicked up by the rumbling traffic evoke the hazy aura of an old sepia-toned photo. Exit the tunnels and you are surrounded by a web of steep streets full of wild Mexican color.

People and vehicles come at you every which way. It's wonderfully old school: bustling sidewalks, chaotic traffic circles, wafts of music drifting out of tiny storefronts. Few straight, flat paths exist in Guanajauto. The city, about 400 miles northwest of Mexico City, is thrust in a river valley between arroyos and canyons. Most streets climb and dip quickly and turn sharply – constantly. Stand almost anywhere and patches of the city unfold below you. Lush greenery rises above.

Color flows like a natural force. Purple bougainvillea tumble over black wrought-iron balconies. The peeling paint of a lime-green wall reveals a previous layer of tangerine. Stop to admire, say, the stone archway of a plaza's entrance, and framed within the columns is a periwinkle wall across the street. Beside that wall is another of deep red.

But there's one color, which is also a precious mineral, that defines Guanajuato: Silver. The Spanish established their first mine in the New World here in the 16th century, and for two centuries the area produced close to 40 percent of the world's silver.

It made the ruling class filthy rich. They lived in haciendas the size of country clubs. They built civic jewels like the University of Guanajuato and quaint plazas and theaters that reminded them of Europe and constructed gaudy Baroque Catholic churches to convert the Indians.

Miners and their families faced abject poverty. Working in the mines, submerged more than a mile underground, tended to be lethal. Little wonder the locals were among the first to rise against the Spanish during Mexico's War of Independence.

Nowadays, Guanajuato mines its rich cultural treasures.

Colonial oppression aside, the Spanish knew how to relax in fabulous style. Even the alleys, called *callejones*, are romantic. Tiny shops and low-slung homes envelope the stone pathways, which often lead to small plazas that feel like wonderful community centers. A typical plaza scene: Families and friends lounge underneath the shade of a tree. Teenagers flirt by a fountain. A guitarist strums a few tunes in a café.

A popular local book, *Leyendas de Guanajuato*, tells of legends connected to 42 streets, *callejones* and plazas. Among the most famous is Alley of the Kiss, or *Callejón del Beso*, a tale of star-crossed lovers, a jealous, violent father and a faithful servant. On any given night, Renaissance-era pied pipers in black tights and puffy coats – *estudiantinas* – recount this story to willing crowds.

CROOKED, STEEP PATHS OF LOVE AND LUST.
Estudiantina groups have a long lineage. Their costumes reflect their ancestry:
College boys who sing and perform street theater for money and to woo women.
Guanajuato is considered the birthplace of estudiantinas in the Americas. They still
lure tourists and others through the heart of the city to the Alley of the Kiss,
or Callejón del Beso (opposite page, inset).

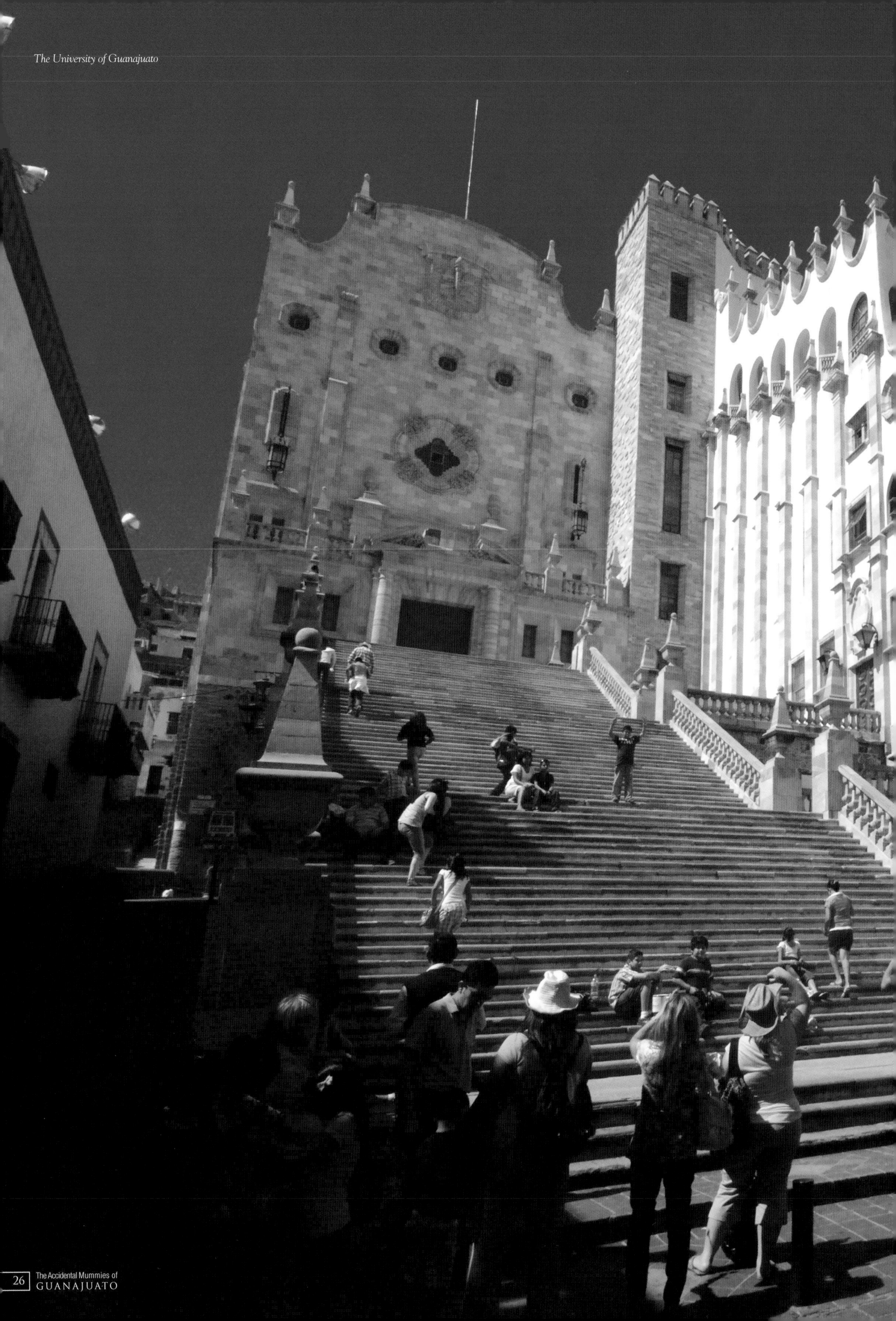

The University of Guanajuato

The University of Guanajuato keeps pumping fresh energy into the city. The city also is the state capital. Both influences supply an intellectual depth. On any given week, it's not unusual to find classical musical performances, foreign films, a jazz combo or some kind of arts festival.

And for three weeks every October, the International Cervantino Festival transforms the entire city into a living theater. The festival is named in honor of Spanish author Miguel de Cervantes, best known for the novel *Don Quixote*. More than 2,000 artists, representing every medium imaginable and from more than two dozen countries, converge here. Indigenous influences are everywhere.

The many fine artisans imbue it in their work and so many street festivals and traditions are rooted in non-European culture. In 1998, the city was named a World Heritage Site by UNESCO. Only a handful of cities in the world hold the distinction and among the criteria is "to exhibit an important interchange of human values." In Guanajuato, that interchange bends and curves with the streets and *arroyos*. It blends centuries and cultures into a modern, creative city. ᘛ

PLACE OF FROGS:

Guanajuato exudes history. The Purépecha Indians, also called Tarascan Indians, named the mountainous region "Quanax-juato": "place of frogs." The Spanish first arrived in the region in 1522. Three decades later, Captain Juan de Jaso discovered something beneath the surface that is the stuff of colonial dreams: Vast amounts of silver. It created a city with few rivals anywhere in the colonies. Bullet holes still pock some city buildings from the early 19th century battles that saw Mexico break from Spain. Today, the city is as Spanish as any in Mexico and yet stubbornly its own creation.

GORKY GONZÁLEZ *is one of Mexico's most celebrated master artisans. Like all of Guanajuato, he seems to blend many influences with ease. His finely glazed pottery pieces are sought by the country's leading architects and interior designers. Gorky, named after Russian writer Maxim Gorky, also has an international following.*

The ceramist revived a technique called majolica, which had been abandoned in Mexico in the 1820's after independence from Spain because it was identified with the colonial rulers.

He works with clay from the nearby Sierra de Santa Rosa, a source for Indian ceramists long before the arrival of the Spaniards in the 1500's. Shaped on potter's wheels and dried in adobe-walled storerooms, the pieces are decorated with plant, animal or folkloric figures in subtle blue, green, yellow and beige hues, then baked in modern electric ovens. A tin oxide enamel glaze is applied, and the works are baked again, this time at much higher temperatures. His pottery is featured in the Accidental Mummies exhibit.

THE FEAST OF COLOR:

Color is the great equalizer in Mexico, one of the many indigenous influences very much alive in Guanajuato.
From the most elaborate upscale party to common everyday scenes, the city is a Technicolor dream.

RICH HISTORY ON A GAUNAJUATO STREET.
On any given weekend there seems to be celebration by a neighborhood, club or church.
This weekend scene exhibits many of the playful attitudes toward death and aging.
Note the mummy mask behind the devil. On the left is a young person wearing a mask of
Don Quixote. The city holds an annual Cervantino Festival, named after the Don Quixote
author Miguel de Cervantes, that's become an international arts event.

HIPSTERVILLE:
The University of Guanajuato and historic gems like Teatro Juarez keep the arts and cultural scene flowing in the city today. The young and educated flock to the city.

SUBTERRANEAN HUES:

The city has made an odd pact with its geography and its reliance on labyrinthine tunnels is just one example. Guanajuato may be more than a mile high above sea level, but it depends on underground warrens. The city was originally built over the Guanajuato River, which flowed through the tunnels. For years, buildings were elevated to accommodate repeated flooding. In the mid-20th century, engineers built a dam and redirected the river into underground caverns. The tunnels became major arteries for pedestrians and auto traffic. They conjure up the image of tunnels in mines.

MINES:

Miners still seek veins of silver, gold and more prosaic metals such as iron and zinc today near Guanajuato. The work remains severe. The mines are often a mile below the surface and it's an eerie, dark world of heavy machinery, dusty air and tough men who pray a lot. It's easy to understand why miner Juan José de los Reyes Martínez, "El Pípila," burned down the door of the Spanish stronghold Alhondiga de Granaditas on September 28, 1810, during the first battle in the Mexican War of Independence. A huge statue of El Pípila, right, stands in a prominent place in the city.

R.I.P.

"No real estate is permanently valuable but the grave."

– MARK TWAIN

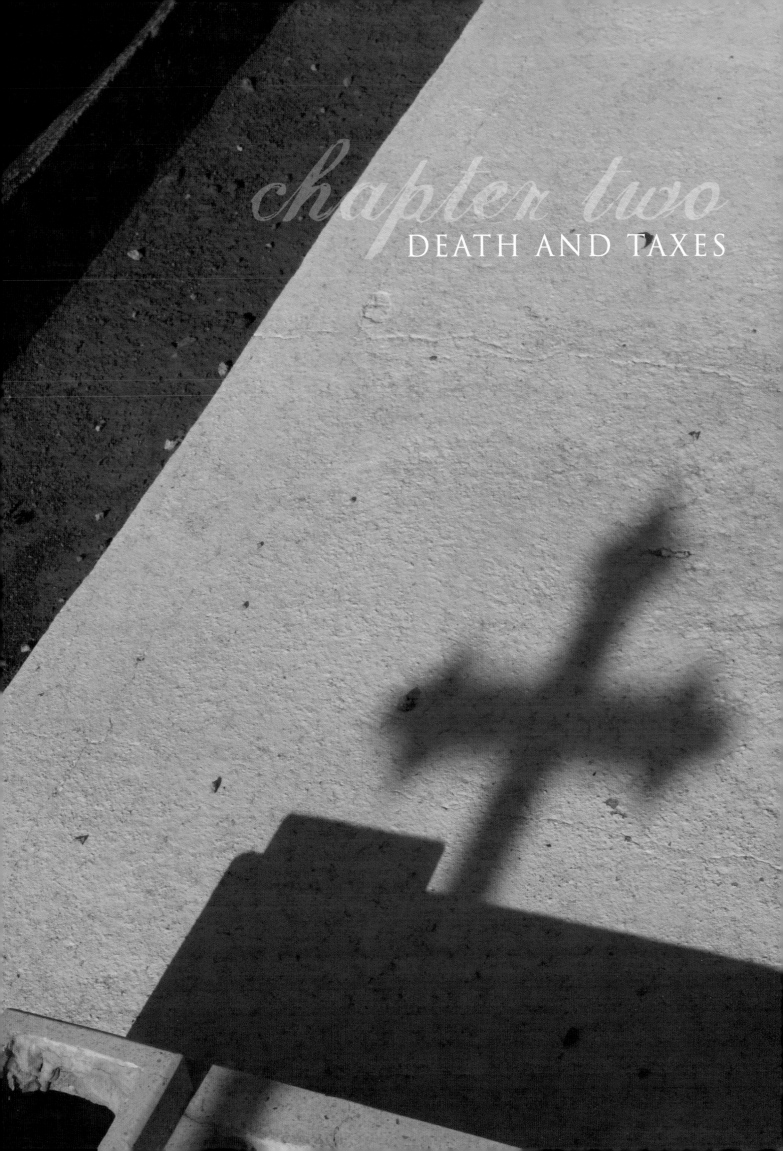

chapter two
DEATH AND TAXES

Guanajuato's graveyard tax was a simple fee done for a basic reason: Someone needs to pay for the space a grave takes up in a public cemetery.

Space plays out in the same wild way at *Panteón Municipal Santa Paula* as it does throughout Guanajuato. There's no room for the old graveyard to grow because it's squeezed between a hill and poor neighborhoods, which sprang up around it over the years. An old beggar or dirty stray dog often linger near the big spooky entrance – a creaky black wrought iron gate and a row of stone skulls that appear to be laughing. Inside the cemetery, the scent of flowers tinges the air and the grounds are thick with greenery and faded white headstones.

The overall layout has abrupt angles and cloistered corners. Long walls of tombs, with seven coffins stacked on each other, form the outer border. During the early days, many of the ruling class preferred to be buried in the walls, which lord over the earth that holds everyone else. Then it became trendy to be in the ground, according to cemetery caretakers. Thus, there are grandiose headstones, and statues of angels and saints, marking the burial sites of generals and people with long titles. They demand so much space they nearly run over the poor, whose graves of tiny crosses with handwritten epitaphs are often just a few inches away.

Space became so valuable it helped lead to the grave tax in 1865. The fee was nominal and one that families could pay over the course of three years if necessary. It's a tax not uncommon in various parts of the world even today. Guanajuato's death levy was enforced for 93 years until 1958.

If the tax went unpaid, coffins were uprooted and the grave sites used again. In the normal course of events, what is removed are skeletons.

At Santa Paula, the natural course of events did not occur 112 times. It is from the concrete wall of tombs, where coffins were sealed in near airtight conditions and bodies lay untainted by even the rare moisture of rain or groundwater, that something more than skeletons have been discovered.

They resemble those rough drafts of a sculptor; the first tendons of clay framed into muscle and patches of brown skin. Unpredictable fragments of clothing remain on some and many women had much of their hair. Any drop of liquid in them had long ago dried up. All of them had open mouths as if frozen in a scream.

The phenomenon struck both rich and poor, young and old. They are miners and infants, doctors and revolutionaries, murder victims and one possible witch.

They are the accidental mummies of Guanajuato. ❧

FRANCISCO
DE P.
CASTAÑEDA

DICIEMBRE 16
DE 1898.

R.I.P.

"The word death is not pronounced in New York, in Paris, in London,
because it burns the lips. The Mexican, in contrast, is familiar with death,
jokes about it, caresses it, sleeps with it, celebrates it, it is one of his favorite
toys and his most steadfast love."

– Octavio Paz, Mexican author, winner of 1990 Nobel Prize in literature

IRONY IN FULL BLOOM
The brooding entrance of Panteón Municipal Santa Paula is a towering ash-gray stone bordered by skulls and with a florid black iron gate. Like most of Guanajauto, this 19th century cemetery is grand and humble. The poor and mighty are squeezed together in a tight space.

PERPETUIDAD
LA NIÑA SAN JUANITA
Granados J. Fallesio el
8 de abril del 66 sus padres
y hermanos le dedican este
Recuerdo
17-2-92

SIMON
SOTO
S.

"You must remember your dead sweetly"

— A PARKING LOT VENDOR OUTSIDE THE MUSEUM,
EXPLAINING WHY HE SELLS MUMMY-SHAPED CANDY.

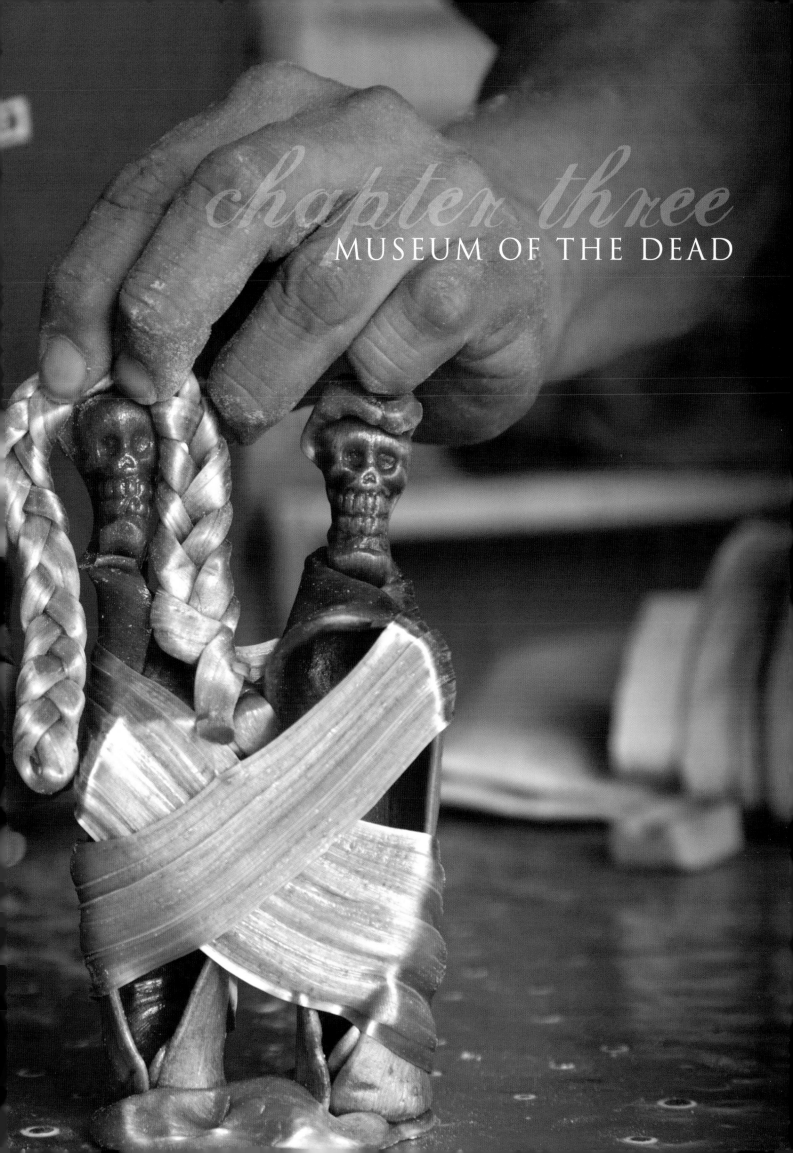

On the morning of June 9, 1865, the remains of Dr. Remigio Leroy were removed from crypt #214 at *Panteón Municipal Santa Paula* because no one could be found to pay his grave tax of 50 pesos.

When the cemetery caretakers pried opened the French physician's wooden casket, located in the middle of a concrete wall of tombs, the men were horrified, according to one of many local legends. One man immediately fled and another fell to his knees in prayer because both feared they had just unleashed the devil.

Dr. Leroy should have been a skeleton. He died three years earlier during a cholera epidemic. His beard appeared to continue to grow beyond his death, but his eyes had vanished. With his dropped lower jaw and his head slightly tilted to the right, Dr. Leroy still looked as if he was in the middle of an engaging conversation.

Was it the hand of God or Satan?

The brightest minds of 19th-century Guanajuato – priests and politicians and philosophers and scientists – gathered to examine him. Even an old Indian woman said to be able to communicate with the dead was brought in to take a look. She allegedly ruled Dr. Leroy wasn't sufficiently dead for her spiritual powers to work.

Little was concluded beyond the fact that Dr. Leroy was actually dead and posed no health threat. He was a mummy, like the ones in Egypt, except that he was an accidental one. The most decisive action came from the cemetery's underpaid caretakers. They started charging admission to the steady stream of curious who wanted to see the mummy. So began a strange new industry – even for a country that joyously celebrates its dead every year. Dr. Leroy was the first of 112 mummies to be discovered. Each was pulled from the virtually airtight wall of tombs, free of bugs and dirt.

By 1894, 29 years after Dr. Leroy was unearthed, the first El Museo de las Momias opened. It was in the same location where the mummies were stored, in catacombs underneath the hilltop cemetery.

Oddly enough, while city officials kept strict records of which dead person owed taxes, other pertinent information, such as the name of the person removed, have somehow been lost.

The lack of real information helps explain why so many feel free to cast whatever belief and story they want on the mummies.

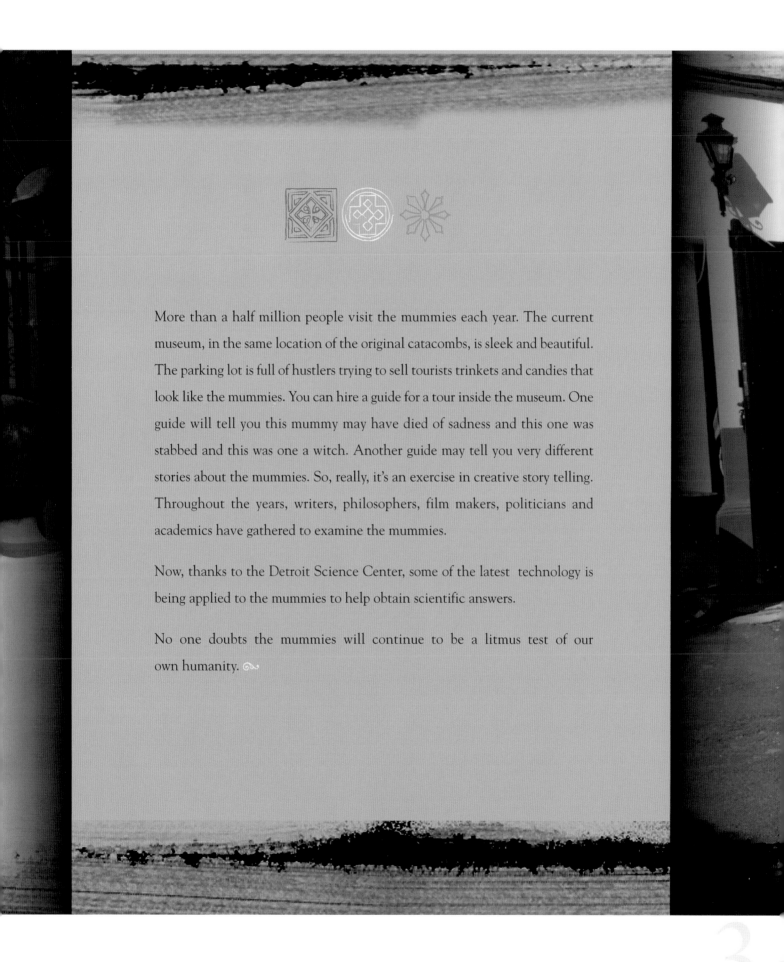

More than a half million people visit the mummies each year. The current museum, in the same location of the original catacombs, is sleek and beautiful. The parking lot is full of hustlers trying to sell tourists trinkets and candies that look like the mummies. You can hire a guide for a tour inside the museum. One guide will tell you this mummy may have died of sadness and this one was stabbed and this was one a witch. Another guide may tell you very different stories about the mummies. So, really, it's an exercise in creative story telling. Throughout the years, writers, philosophers, film makers, politicians and academics have gathered to examine the mummies.

Now, thanks to the Detroit Science Center, some of the latest technology is being applied to the mummies to help obtain scientific answers.

No one doubts the mummies will continue to be a litmus test of our own humanity.

One of the original rooms of El Museo de las Momias.

The Accidental Mummies of
GUANAJUATO

The infants are among the most popular mummies.
The girls are dressed like angels and the boys like saints.

Photos on the wall are of the burial of an infant believed to have become mummified.

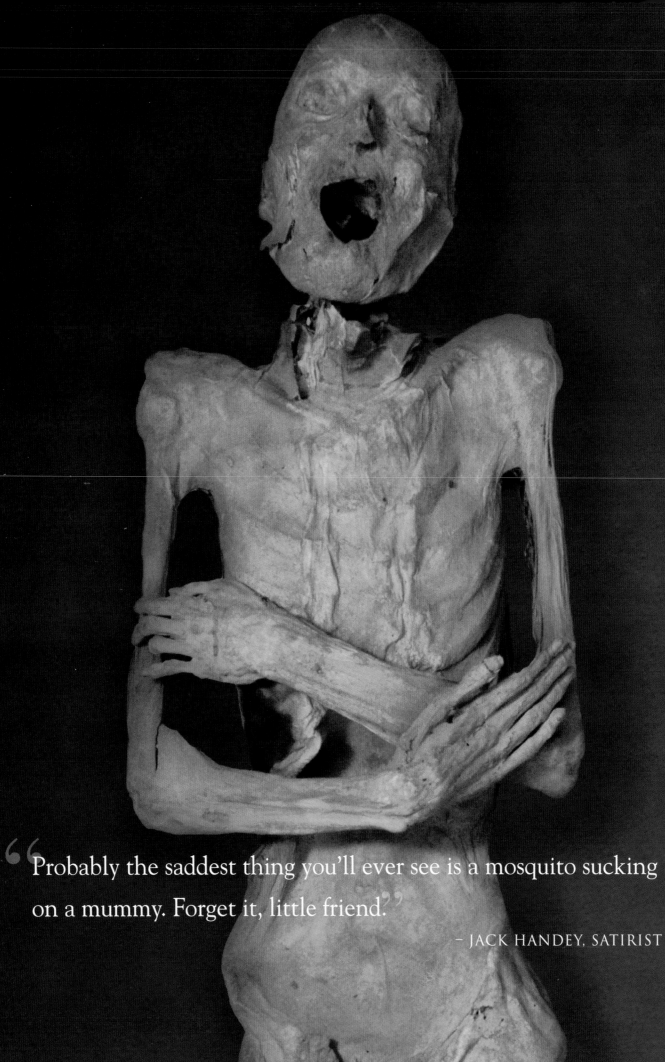

" Probably the saddest thing you'll ever see is a mosquito sucking on a mummy. Forget it, little friend."

— JACK HANDEY, SATIRIST

chapter four
THE FAMOUS DEAD

The naked truth of how we become skeletons:

It starts with our gut. In a matter of days after being buried, bacteria inside our intestines feed on the contents of the intestine and then on the intestine itself. These bacteria break out and invade the body cavity and start to digest other organs. At the same time, enzymes inside each cell seep out, devour the cell and its connections with other cells. Then come the bugs. Flies lay eggs around wounds and other body openings – mouth, nose, eyes, etc. The eggs hatch and larvae – maggots, really – move in to feast on dead tissue. We begin to rot and stink.

We emit gases, including hydrogen sulfide, methane, cadaverine and putrescine. The stench attracts beetles and mites. Our bodies begin to bloat, providing a nice warm living space for the maggots.

All this takes place within three weeks of our burial.

Our bodies collapse and our skin becomes like rotting fruit. In two months time, we begin to dry out. Mold grows on our body as beetles chew through skin and ligaments. Within a year or less, moths and bacteria eat our hair, leaving nothing but bones.

Here is what likely happened at *Panteón Municipal Santa Paula*: All the mummies come from the middle sections of the wall of tombs. Encased in wood coffins and in virtual airtight cement, only the first portion of the process really started. Every drop of liquid and working organ dried up quickly. But that's as far as it went. The rest of their bodies and clothes that normally gets devoured by time have survived, though it's unpredictable what remains actually remain.

Scientists say the Guanajuato mummies represent a wide cross-section of people, which is rare and beneficial from a study perspective. Another plus for scientists is the wide range of time they have been mummified – some more than a century and others a few decades. All of this gives many of the Guanajuato mummies "distinct personalities." All of them still have skin to varying degrees.

Several women still have long braided hair even as bones tear through parts of their decaying bodies. Tattered pieces of clothing often survive, a single stocking, a pair of black pumps – even as moths eat away their abdomens. The naked ones reveal flattened breasts covering their chests like shields. A woman who died while giving birth has a tragic tubular pattern embedded in the hollow folds of her stomach.

Many of the skulls on the men have snapped off from their bodies. Their genitalia usually remain whole. One man thought to be a solider during the Mexican Revolution still wears a uniform. Another dons a black suit and fancy leather shoes but no longer has a nose. One wears only a chin strap, stretching from his lower jaw and tied to the top of his head. X-ray images reveal the cloth from the strap has wedded to his skin and bones.

Most mummies look as if they are about to speak; the result of decaying jawbones. "Theirs was a perpetual screaming," Ray Bradbury writes in his short story, *The Next in Line*. "They were dead and they knew it. In every raw fiber and evaporated organ they knew it." ❧

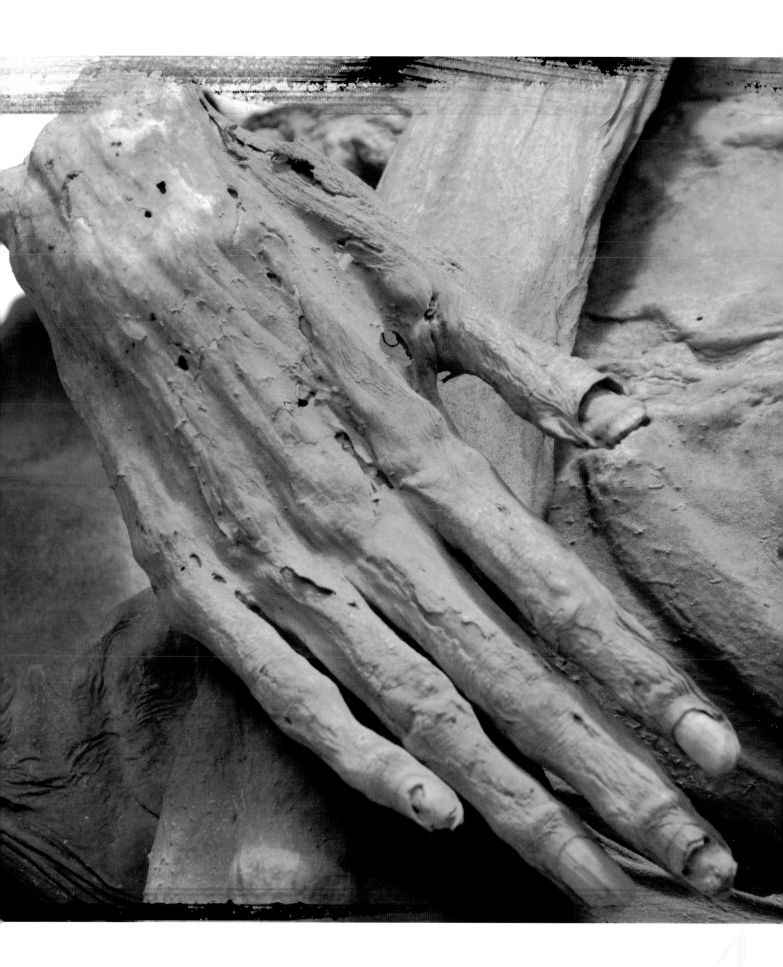

the infants:

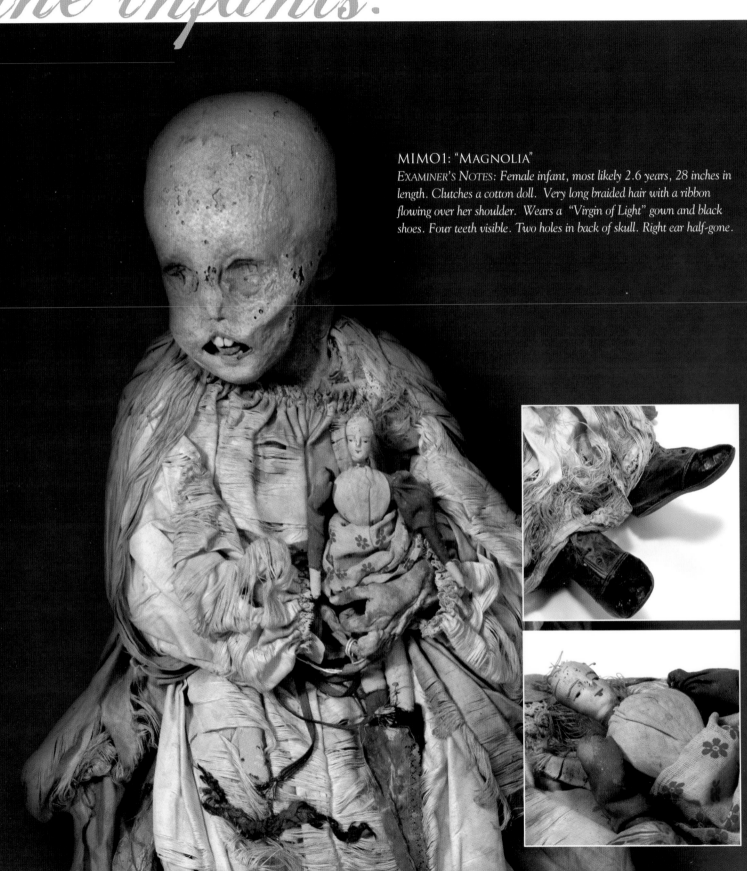

MIMO1: "Magnolia"

EXAMINER'S NOTES: *Female infant, most likely 2.6 years, 28 inches in length. Clutches a cotton doll. Very long braided hair with a ribbon flowing over her shoulder. Wears a "Virgin of Light" gown and black shoes. Four teeth visible. Two holes in back of skull. Right ear half-gone.*

THE COLD HARD FACTS:

The basic information of the mummies reads like a vague crime report. They have been assigned a sterile mix of letters and numbers and often fictional names for categorical purposes. Their physical descriptions are blunt and haunting.

MIMO2: MAGDALENA AGUILAR *(real name)*
EXAMINER'S NOTES: Died Sept. 8, 1897. Entombed in niche #14 of the sub series #6. Eyes "mummified" into a piercing stare. The frilly lace of her baptism gown still intact. Copious larval casings, likely Demestid Beetles, in her open mouth. Exhumed 1909.

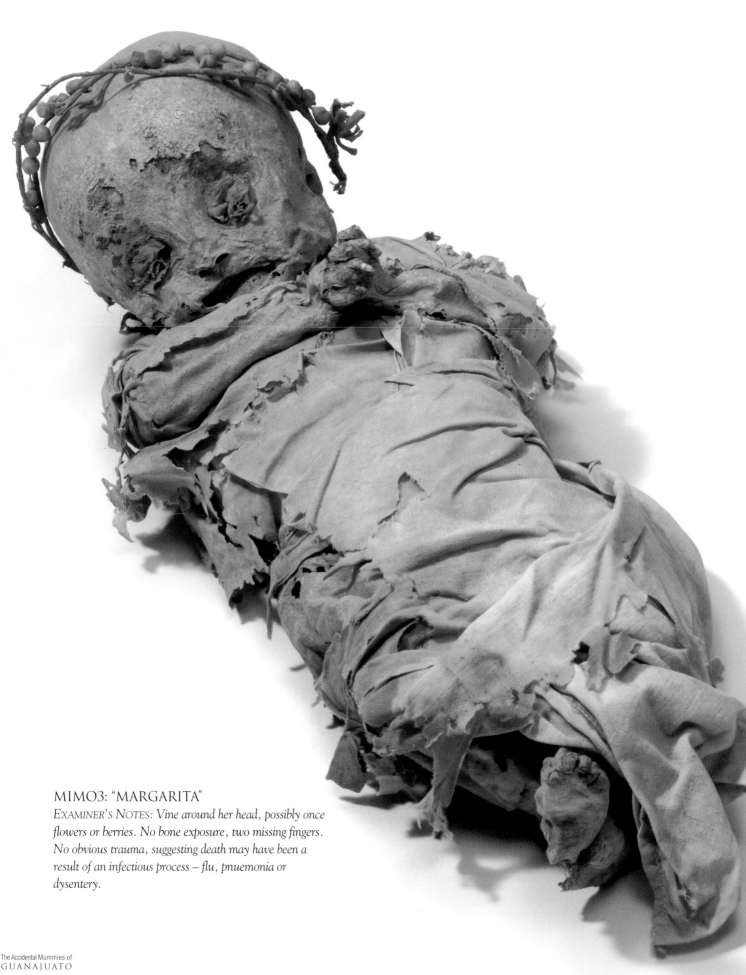

MIMO3: "MARGARITA"

EXAMINER'S NOTES: Vine around her head, possibly once flowers or berries. No bone exposure, two missing fingers. No obvious trauma, suggesting death may have been a result of an infectious process – flu, pnuemonia or dysentery.

MIH07: "MARQUITO"

Examiner's Notes: Boy of unknown age holds heart-shaped pendant. Suit of Sacred Heart of Jesus; deteriorating shoes and filthy socks. Full face. Head loose. Skull split.

the women:

MM27: "LA BRUJA" (THE WITCH)
EXAMINER'S NOTES: *Old adult female. Majority of face destroyed. Left forearm detached. Arms moth-eaten. Teeth showing through fading jawbone. Floral blouse intact.*

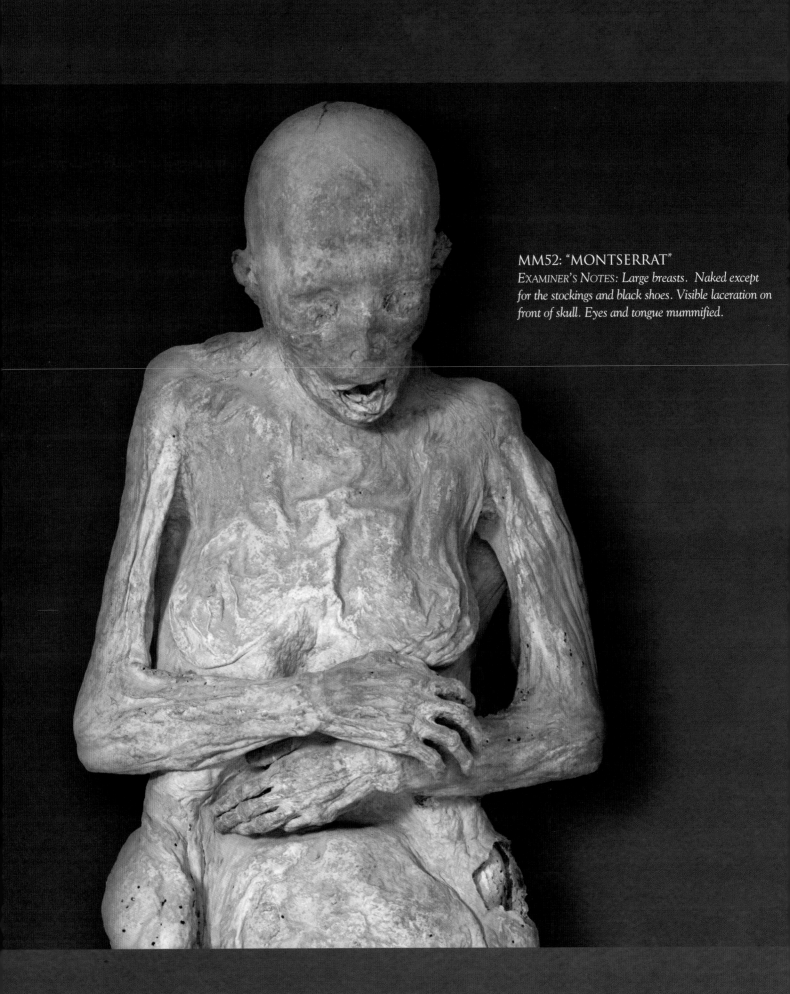

MM52: "MONTSERRAT"

EXAMINER'S NOTES: *Large breasts. Naked except for the stockings and black shoes. Visible laceration on front of skull. Eyes and tongue mummified.*

MM29: "MONICA"
EXAMINER'S NOTES: *Completely nude. Good face.*
Massive tumor visible on right side of abdomen.

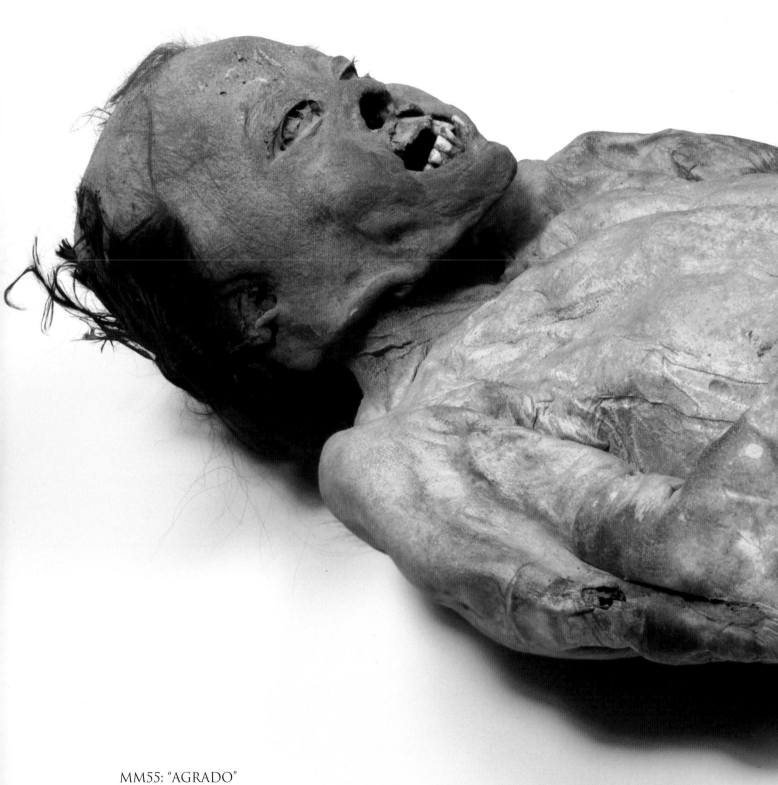

MM55: "AGRADO"

EXAMINER'S NOTES: *Sturdy adult female. Nude. Nose and both ears gone. Skinless hands. Voluminous stomach. Brown stockings.*

MH03: "ARTURO"

Examiner's Notes: Adult male, age between 20 and 30. Large amount of skin missing in the facial area. Three buttons on his jacket display a floral pattern. No ears. Feet tied with a cord. Very nice suit and pants.

MH29: "GREGORIO"

EXAMINER'S NOTES: *Nude adult male with slightly moth-eaten shoulder blades and face above the eyes. Skin of left calf missing. Deep gash on underarm at left shoulder. A small ragged hole in right ribcage. Hole appears to have some type of gauzing inside.*

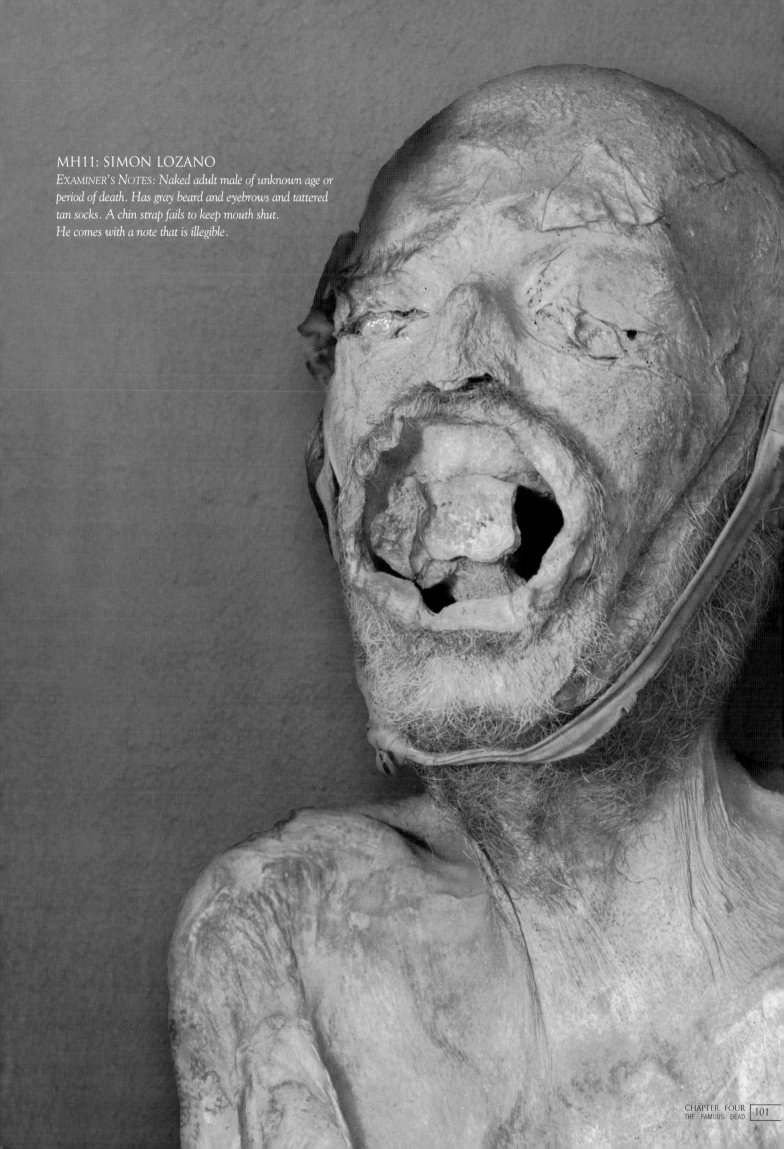

MH11: SIMON LOZANO

Examiner's Notes: Naked adult male of unknown age or period of death. Has gray beard and eyebrows and tattered tan socks. A chin strap fails to keep mouth shut. He comes with a note that is illegible.

MH13: VIVTOR

Examiner's Notes: Adult male believed to be a miner – source of speculation unknown. He wears a shirt, jacket, pants and belt. Mummified eyes and tongue. Scar on cheek and neck.

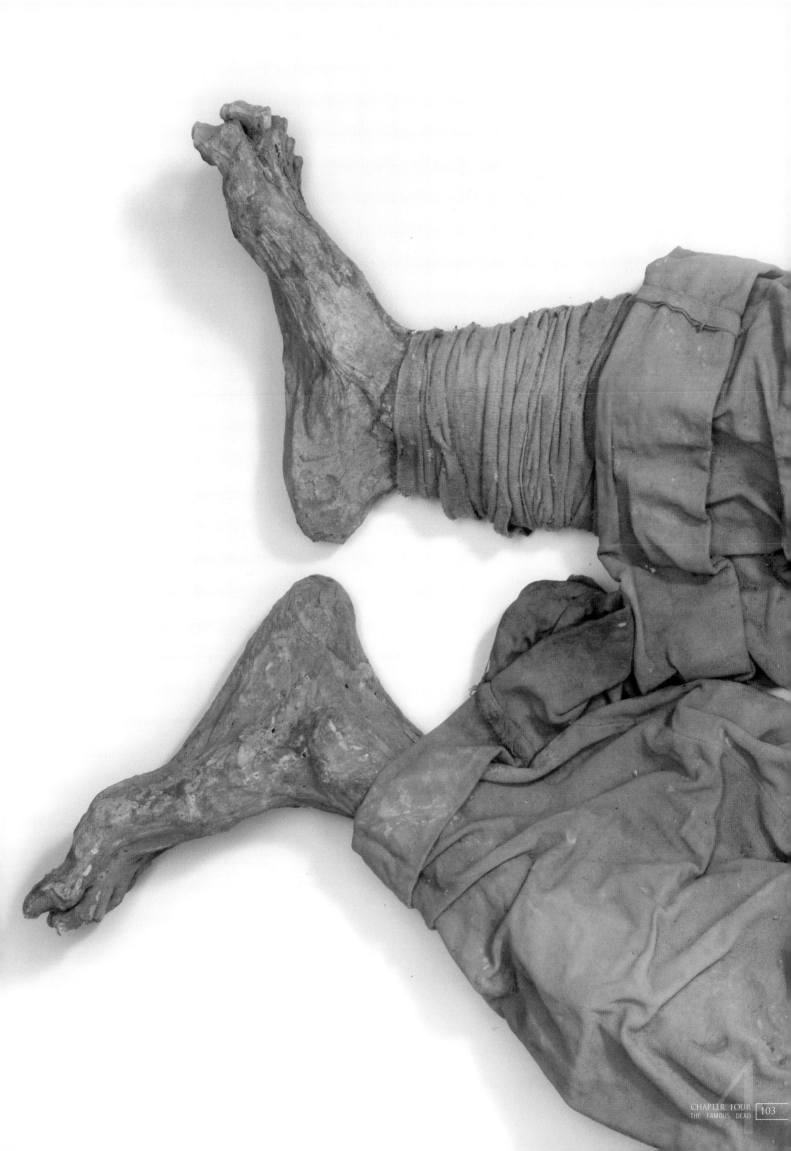

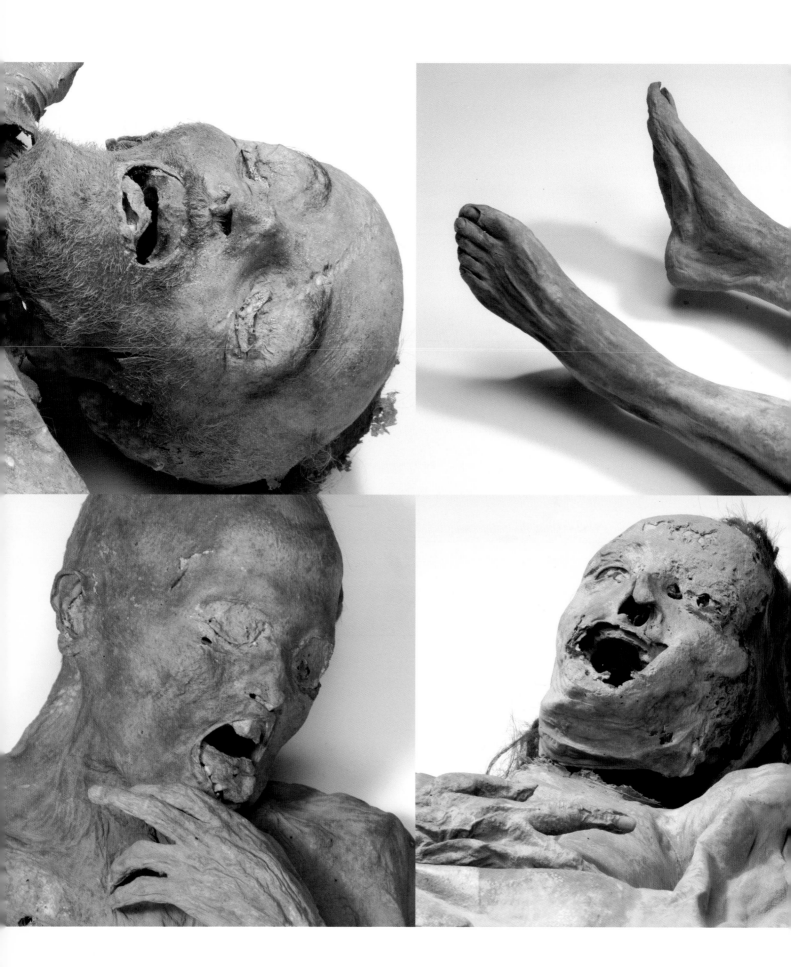

The Accidental Mummies of
GUANAJUATO

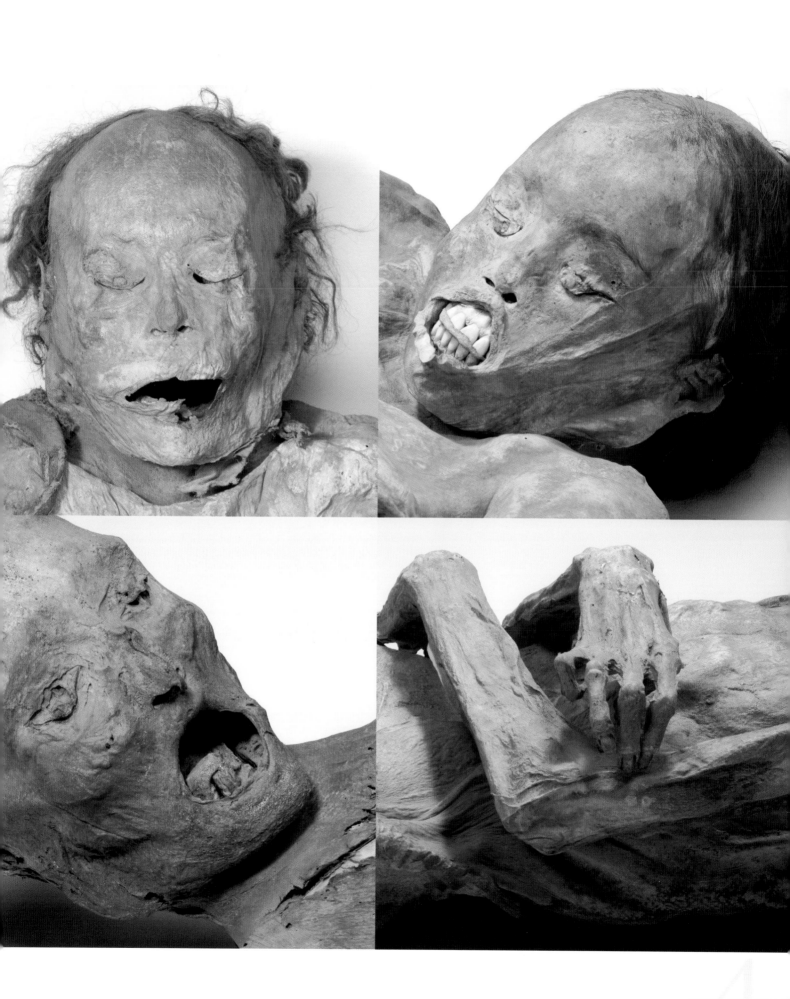

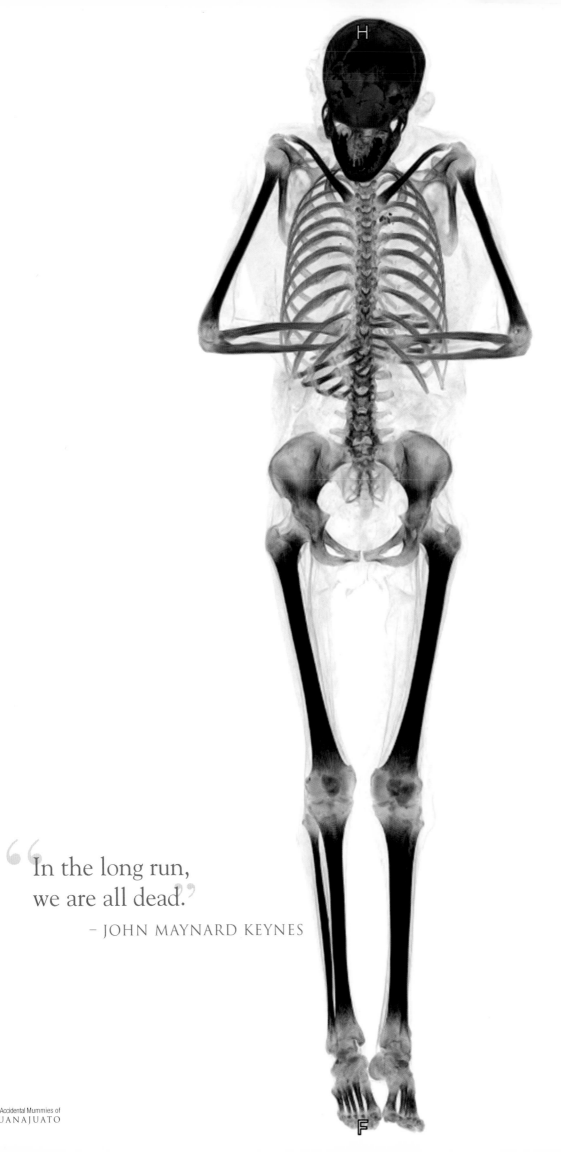

"In the long run,
we are all dead."

– JOHN MAYNARD KEYNES

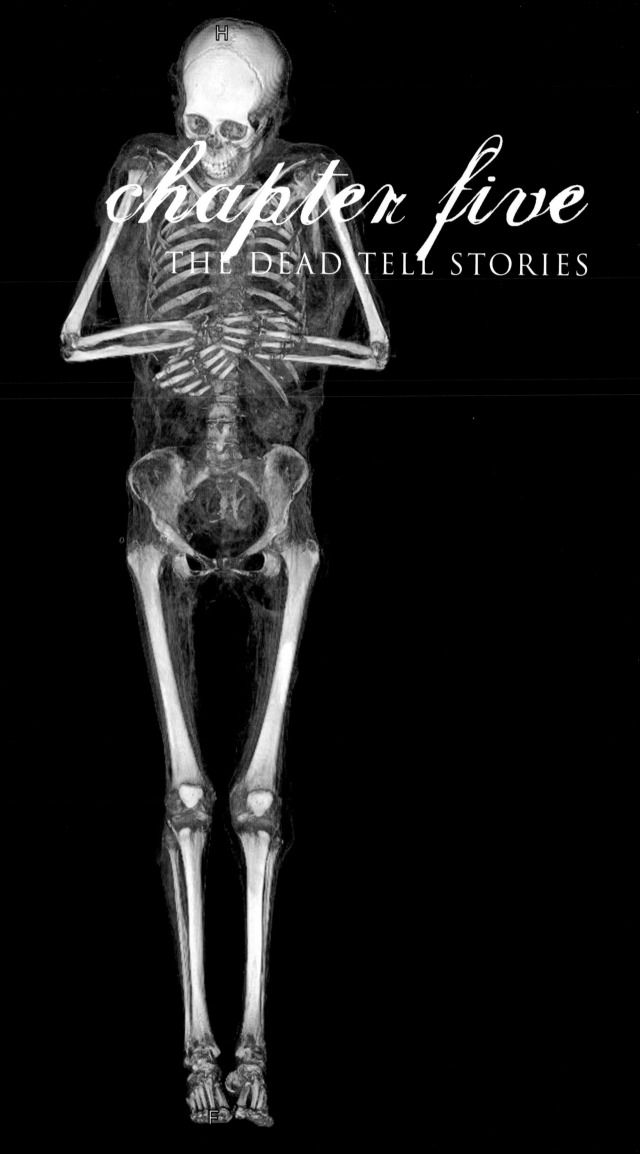

chapter five
THE DEAD TELL STORIES

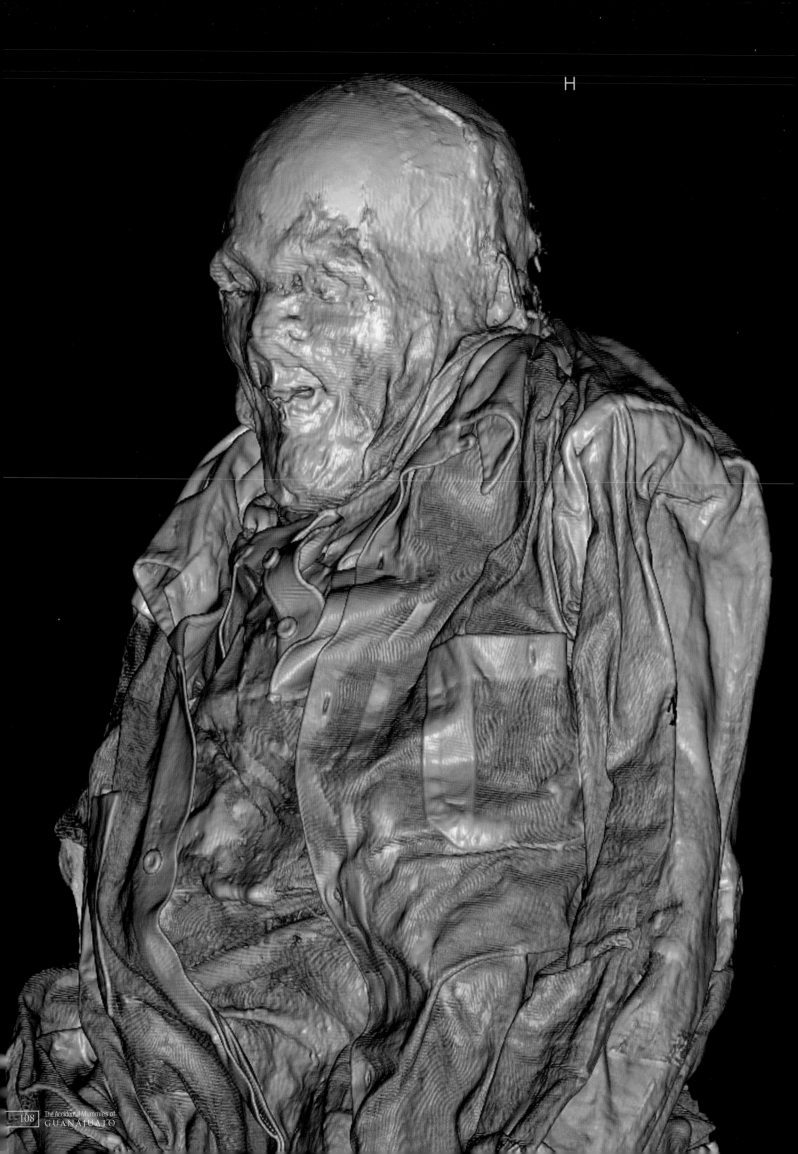

H

In the Philippines, American scientists Ronald Beckett and Jerry Conlogue once agreed to the custom of being honored with the sacrifice of three black pigs, which they witnessed at very close range. Then they ate the pork raw – what exact type of pork, they'd rather not say.

In Peru, they drank ceremonial beer containing human saliva. In Italy, the middle-age professors squeezed inside graves deep in a cave to examine the mummified dead. Mummy devotees like Beckett and Conlogue will go to great lengths to follow their muse of decay. To go where few scientists have gone before.

Probe dehydrated organs and ravaged cells with fiber optics that fit on the head of a pin: **endoscopy**.

Unravel the secrets within disarticulated skulls and absent extremities via electromagnetic radiation: **X-ray**.

Create thousands of precise cross-sectional images of dissipated biology through computed tomography: **CT scans**.

Reconstruct a human face, through clay sculpture and sketches, from a being often viewed as a mere freak of nature: **forensic art**.

All of this is at the heart of the Accidental Mummies exhibit created by the Detroit Science Center: To apply advanced science to dusty mummies and be part of the ongoing effort to restore their humanity.

It plays out like a detective novel or *CSI* episode. The case: Identify the dead. If not their names, at least, the basic facts: How did they die? How did they live? Find out their approximate age.

It begins in a beautiful old city in Mexico, where the earth once produced amazing amounts of silver and where many of the dead didn't quite return to dust. Those mummies are on display now in a pretty little museum on a hill, where the curious come from thousands of miles to look at them. People have been telling stories about some of them for more than 100 years.

The truth is, however, they may just be stories.

The conscientious museum director, Juan Manuel Guerrero, yearns for more. He and a U.S.-trained physician, Dr. Eduardo Romero Hicks, who became mayor of Guanajuato, invited Western scientists to examine the mummies in the early 21st Century. It was something of a bold move. It wasn't so long ago that scientists in general were viewed as the bad guys – because they often were. For decades, Western academics pillaged native cemeteries for bones and mummies and hauled them to colleges and museums. The law eventually stepped in. In 1990, the United States passed the Native American Graves Protection and Repatriation Act, protecting native graves and ordering many museums to return bones and mummies to native bands. It sent a chill throughout the world.

"We have to stop thinking in a totally Euro-centric way if we are ever going to advance," said Conlogue.

That's why he and Beckett don't mind adhering to local customs to see their mummies; customs like sacrificing black pigs or drinking beer made of spit. They once co-hosted the National Geographic Channel's *The Mummy Road Show*, during which they first encountered the Guanajuato mummies in 2007. The great thing about the Guanajuato mummies is that they represent a wide range of sociological positions and ages, the duo explain.

"Their stories are just too big," Beckett explains. "They are too valuable to ignore."

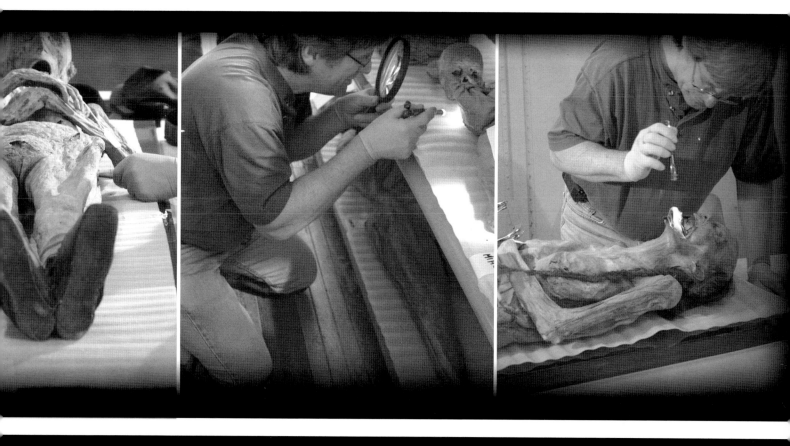

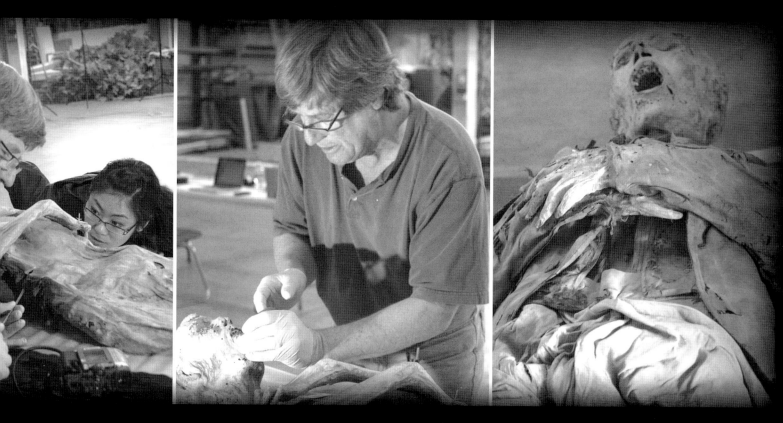

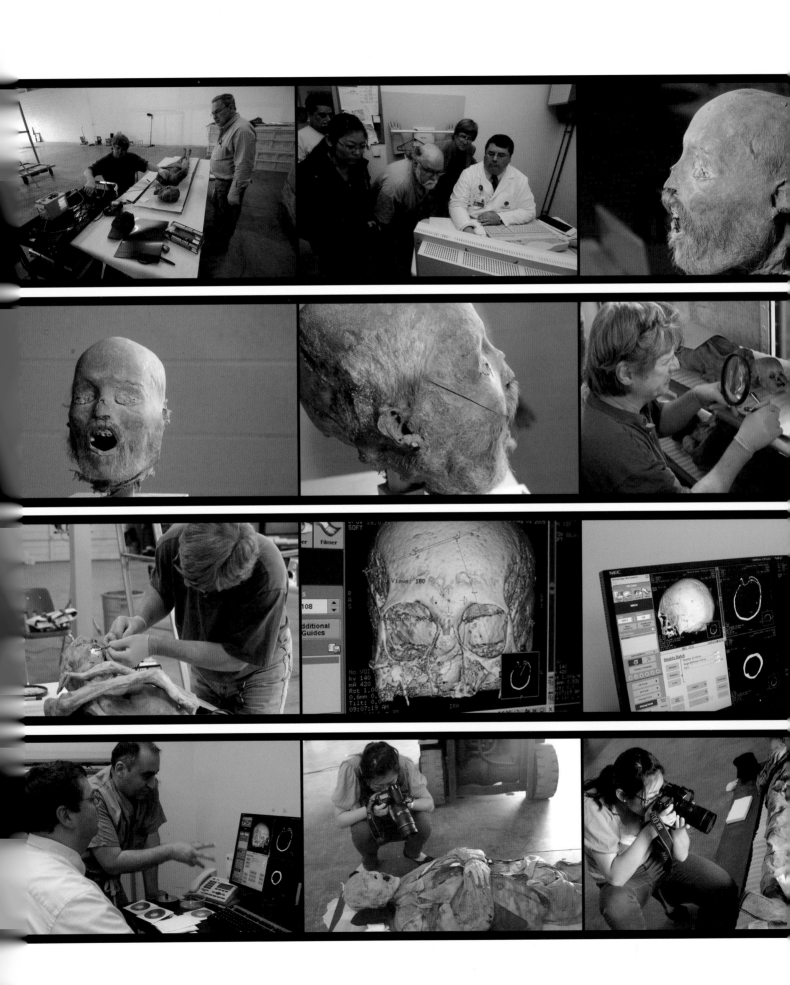

forensic facts:

■ Natural mummification happens all over the world and almost always in extreme climates, where dry sand or freezing cold stop the natural process of decay.

■ Scientists investigating mummies collect data starting with what they can see and touch, which is called gross examination. Measurements are taken from head to toe. Photographs are taken to detail the condition of skin, teeth, bones, nails, hair, clothing and accessories.

■ Bones often survive the process of decay and provide the "primary body of evidence" for determining age, gender, race and possible cause of death.

■ Teeth are the most stable of all bodily materials. Scientists investigate how teeth are worn down, and which teeth are decaying or missing. Teeth can lead to facts about diet, age and health.

■ Trained in human anatomy, archeology, osteology (bone study) and detective work, forensic anthropologists are key players in crime scene investigations, CSI.

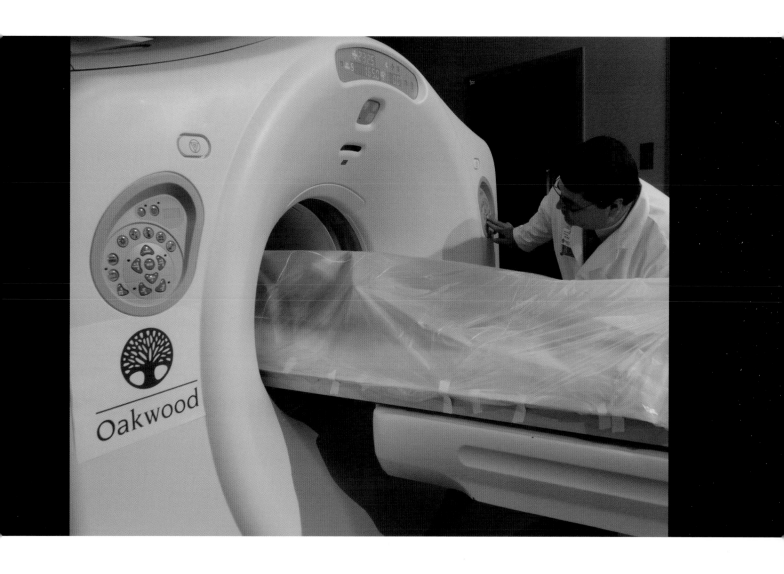

CASE STUDY:
MM27: 'LA BRUJA'

She has been called the Witch (La Bruja) for as long as anyone can remember, and time has eroded the line between truth and fiction about her life.

She looks fierce. Her light hair is elaborately braided and pinned, yet her skull is disintegrating and her mouth is locked in a scream. Her fingernails appear to still be painted. A faded leather medallion hangs around her neck. She looks hunched.

Some believe she was a good witch, a practitioner of herbal medicine and home remedies in real life. Or maybe her striking looks as mummy led many to believe she was a witch.

"It's a shame to just call her a witch," said Jerry Conlogue, the academic who specializes in x-raying mummies. "We'd really like to know something more about her."

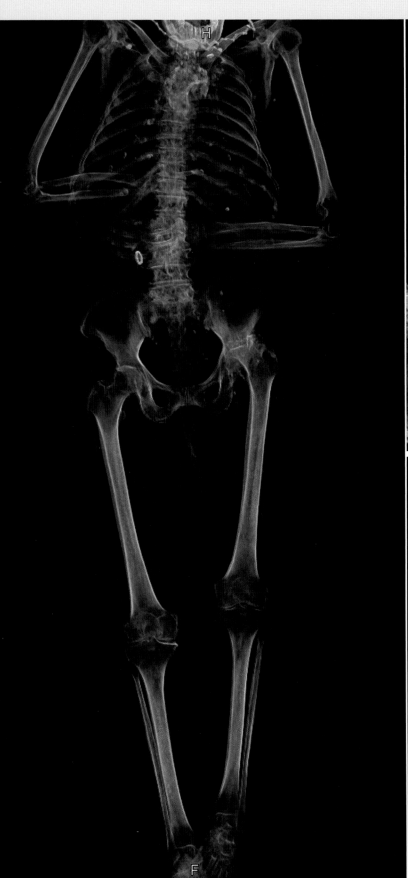

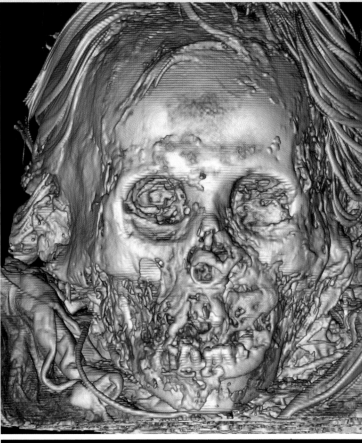

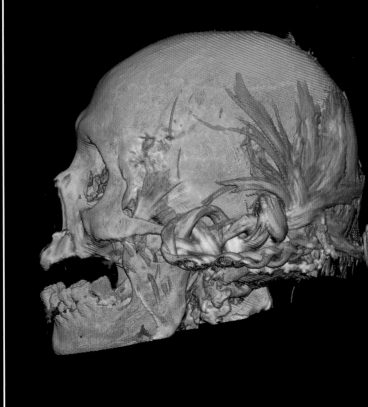

SCIENCE PROVIDES THE ANSWERS

La Bruja was among the mummies given full-body CT scans at Oakwood Imaging Services in Dearborn, Mich.

Through computed tomography, thousands of X-rays are systematically manipulated and colorized to produce startlingly clear pictures of decomposed soft tissue and bone that lie beneath her visible mummified remains.

The radiologists and mummy doctors Beckett and Conlogue winced in mock pain when they saw the CT scans of the woman. It revealed a marked curvature of the spine and severe osteoarthritis in the hips. It means she would have difficulty walking and discomfort in many daily situations.

"This was a woman who was in a lot physical pain in her real life. Maybe it was her appearance that led people to believe she was a witch?" Beckett asks.

Her hair revealed high levels of iron, lead, sulfur, tin and mercury – the legacy of silver toxicity and living in a mining town.

La Bruja, as reconstructed by Barbara Martin Bailey and Christine Chambers

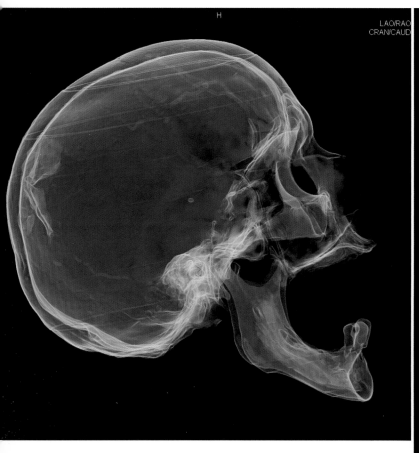

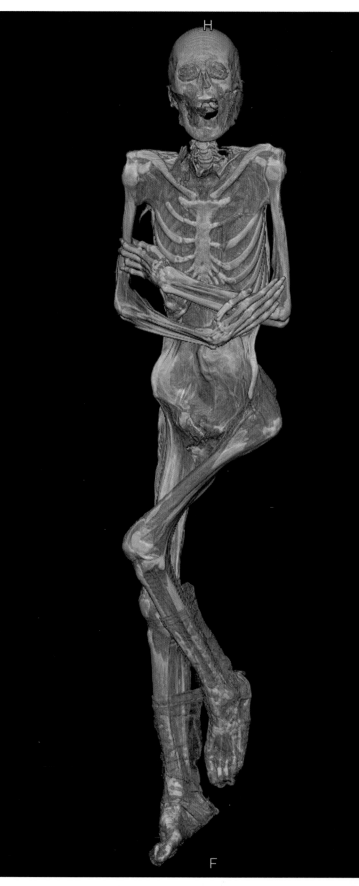

MH28: 'THE MINER'

Consider the Miner's swagger, the curious angle of his hip. What physical trauma or accident in life caused his deformity? X-rays and CT scans reveal what appears to be an old fracture of the hip and a deformity of the knee.

In death, the brain typically liquefies. Excessively dry environments, however, can slow the process of dissolution. Above in the CT study of the Miner's skull we can detect the ephemeral remnants of the brain in the posterior fossa.

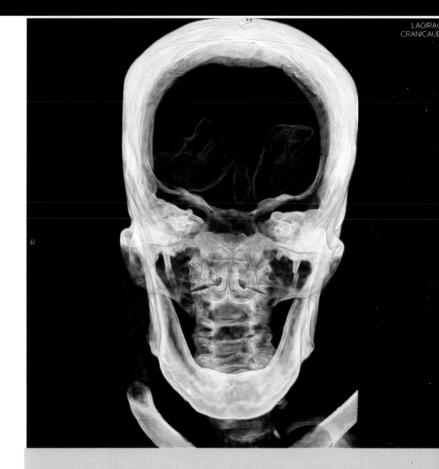

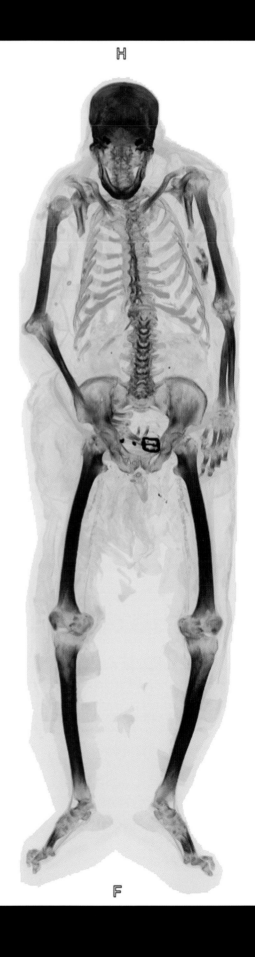

H

F

R

MH13: 'VIVTOR'

CT scans reveal brain remnants and only hints of other preserved organs. Abnormalities of the spine and heels, characterized by extreme formation of bone spurs, are thought to be diffuse idiopathic skeletal hyperostosis (DIS) – a form of arthritis associated with inflammation, stiffness and pain.

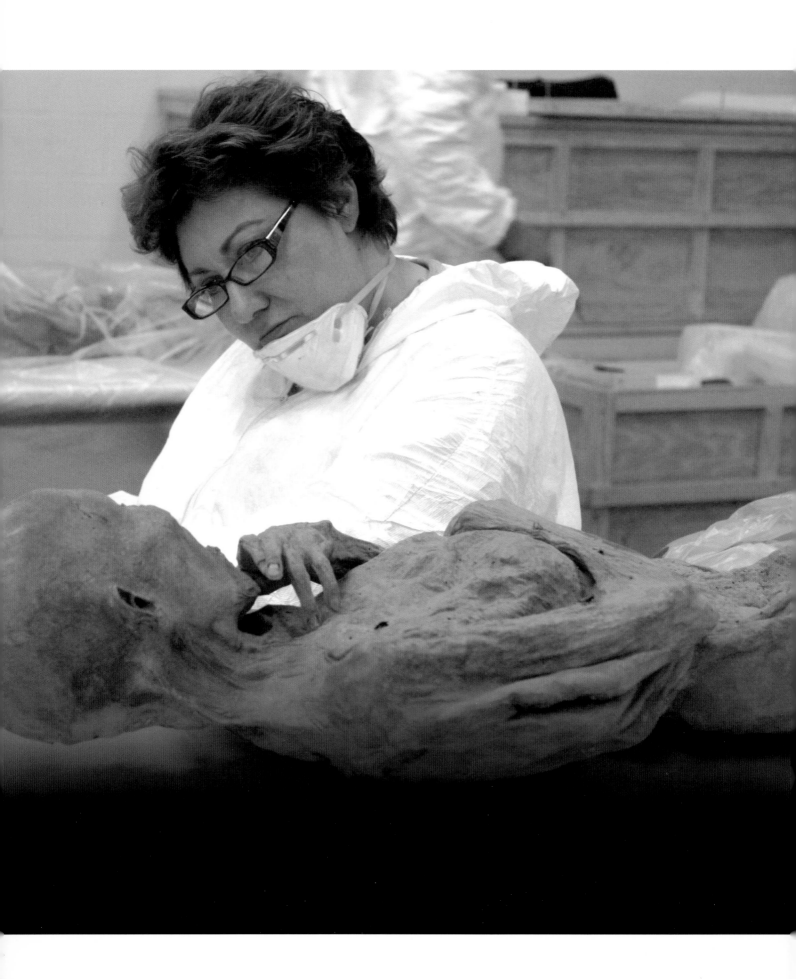

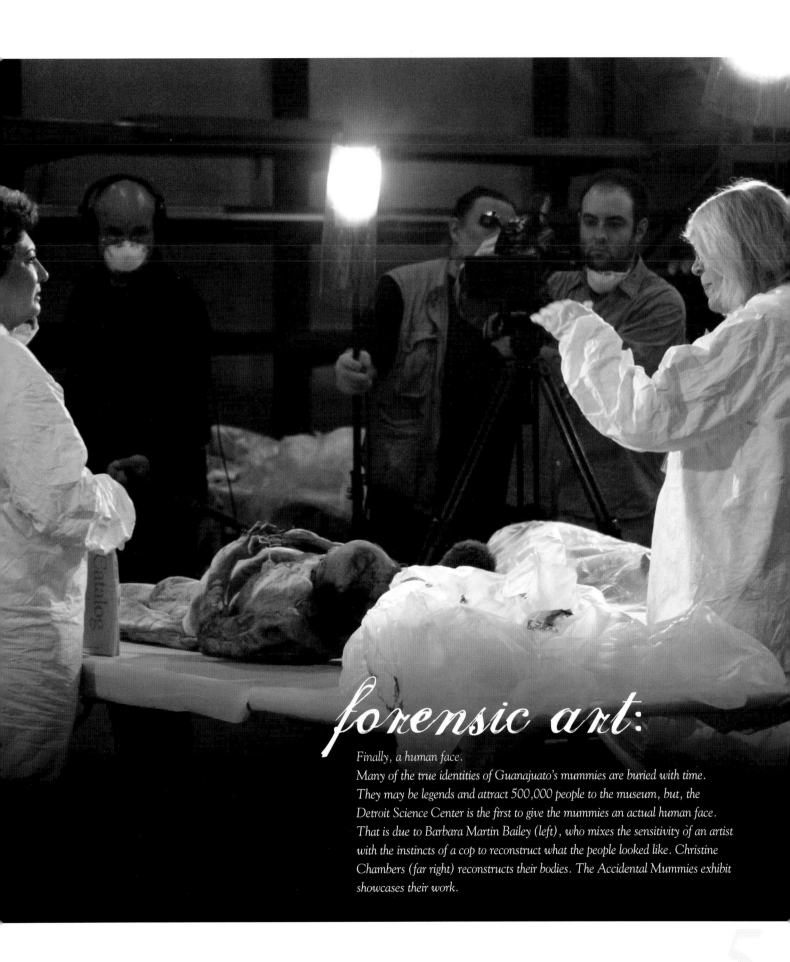

forensic art:

Finally, a human face.
Many of the true identities of Guanajuato's mummies are buried with time.
They may be legends and attract 500,000 people to the museum, but, the
Detroit Science Center is the first to give the mummies an actual human face.
That is due to Barbara Martin Bailey (left), who mixes the sensitivity of an artist
with the instincts of a cop to reconstruct what the people looked like. Christine
Chambers (far right) reconstructs their bodies. The Accidental Mummies exhibit
showcases their work.

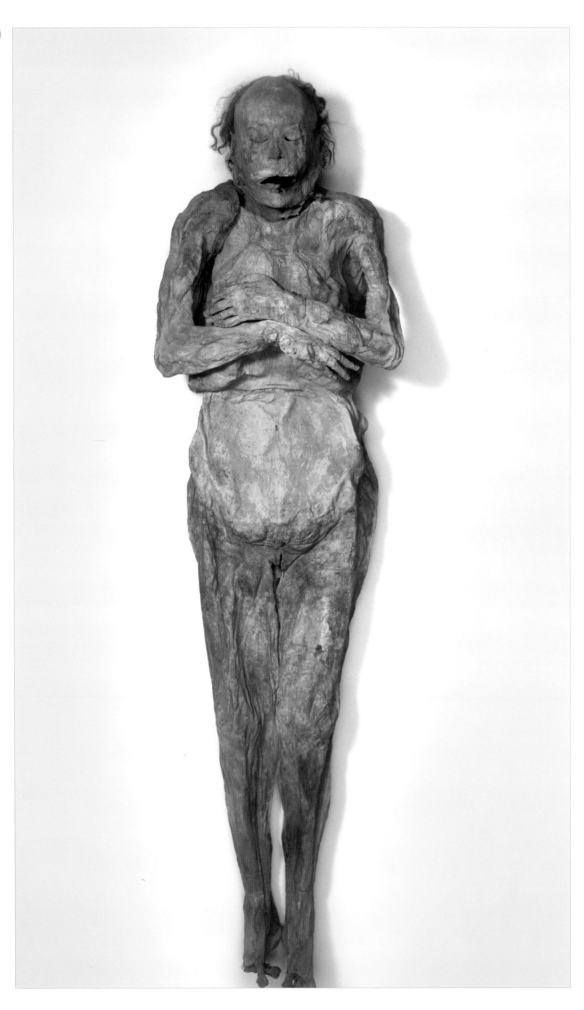

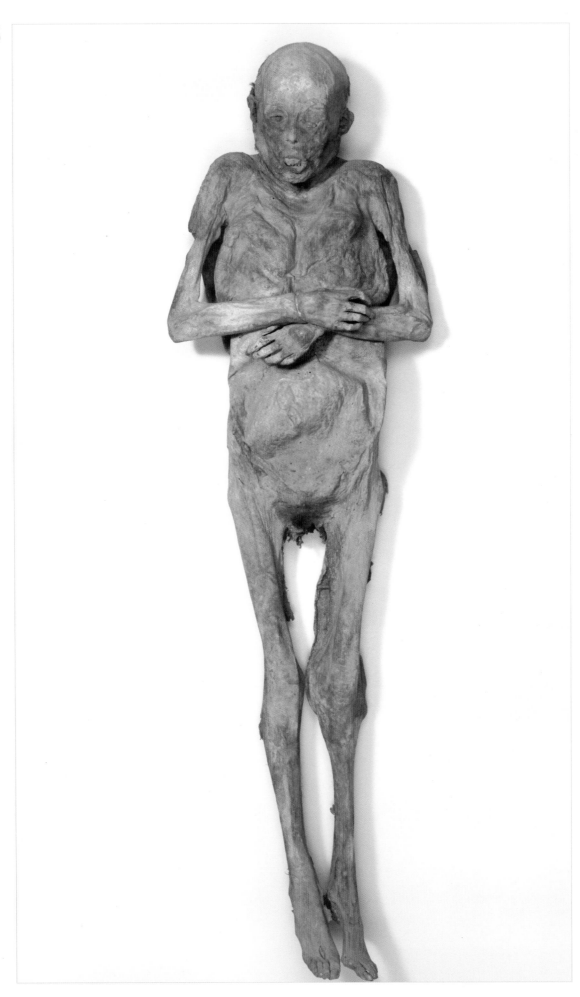

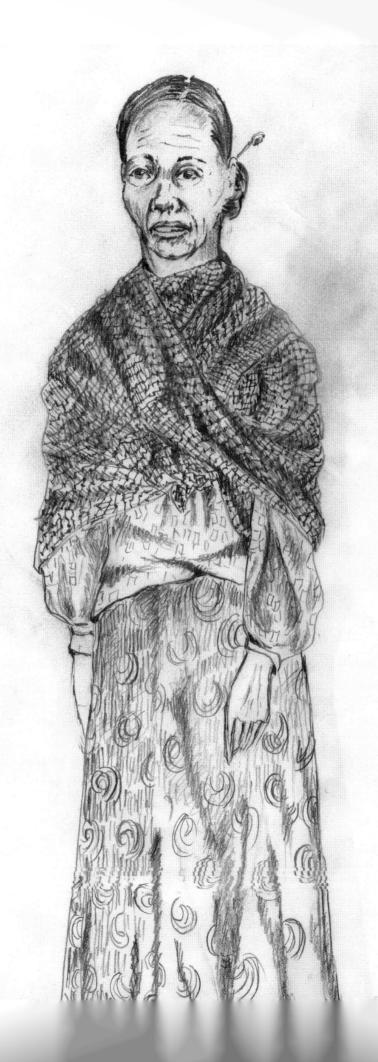

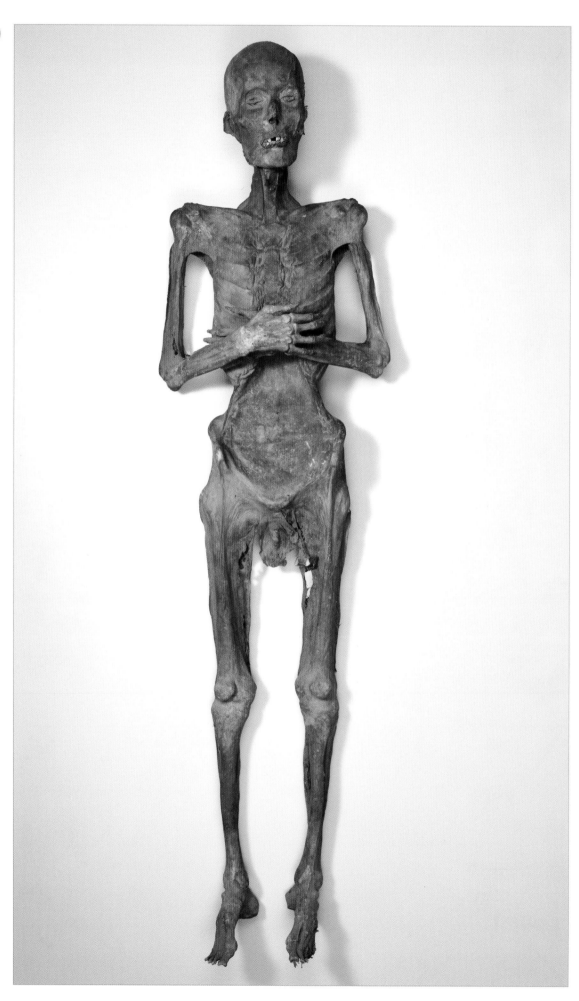

"Our death is not an end if we can live on in our children and the younger generation. For they are us; our bodies are only wilted leaves on the tree of life."

– ALBERT EINSTEIN

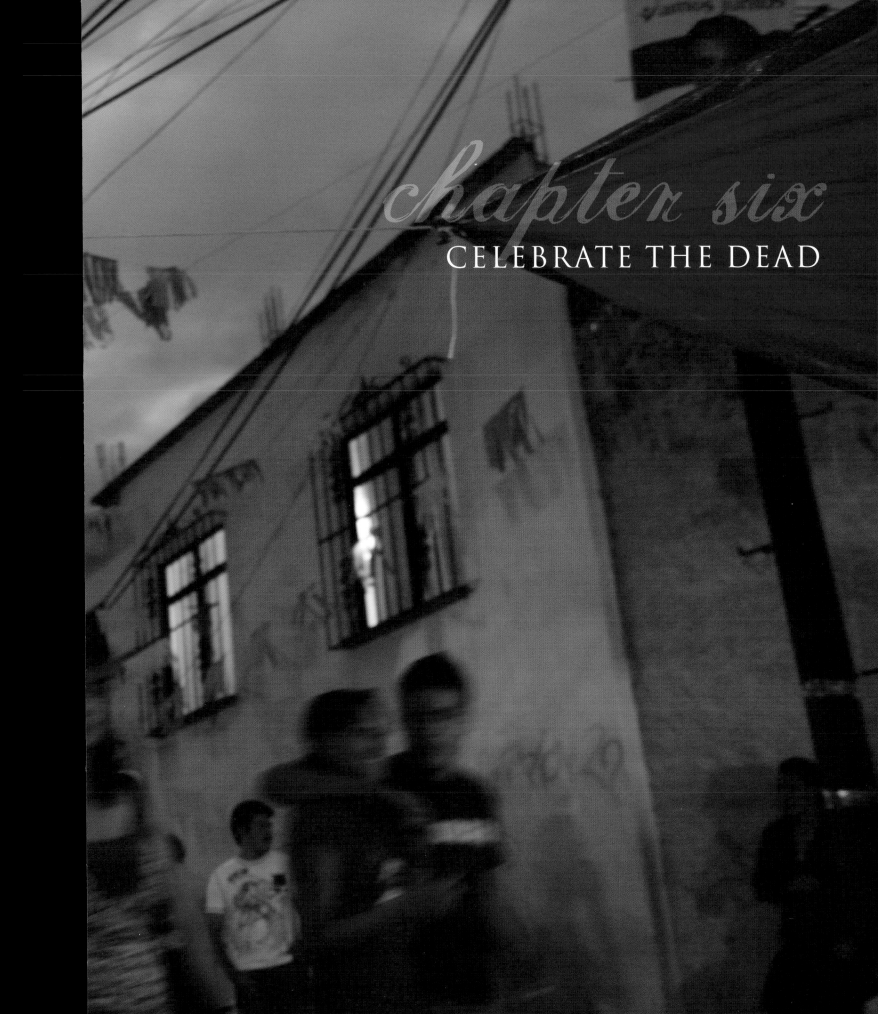

chapter six

CELEBRATE THE DEAD

El Día de los Muertos in Mexico is a time to welcome the souls of the dead. A celebration where the living and the dead are joined. Days before, marketplaces are filled with special toys made of paper mache, candies and paper cutouts, all shaped liked skeletons. Special bread is prepared. Black, purple, white, pink, yellow and golds are main colors. Flowers are vital, especially orange marigolds. Baby's breath, coxcombs, white amaryllis and wild purple orchids called flower of the souls are also prevalent

Day of the Dead is more than a Mexican version of Halloween. It's about the big 'Why.' Why are we here? How will we remember the people who have moved on and what impact did they have in our lives? What will be our legacy?

The essence of Day of the Dead is to honestly deal with mortality: To embrace the basic fact that life and death are natural cycles in this world. The celebrations are neither wild parties nor somber odes.

At their best, they are opportunities to remember the people in our lives who are no longer physically here. Feel sad, fondness, respect, anger – possibly a mix of them all, but allow yourself to feel something so you can get on with life. As far back as 3,000 years ago, many of the indigenous people of what is now Mexico – Aztec, Maya, Mixtec, Olmec, P'urhepecha, Totonac and Zapotec – all held Day of the Dead-type celebrations.

When the Spanish arrived, they sought to kill these festivities. But the best they could do was cover it with a Roman Catholic cloak. They moved the celebrations to coincide with All Souls Day, the Catholic holiday of honoring the dead. In most regions of Mexico, November 1 honors children and infants; adults are honored the following day. It's not surprising the Guanajuato mummies blend into this rich heritage of not shying away from death or its imagery.

In the 20th century, political cartoonist José Guadalupe Posada, 1852–1913, had a lot to do with the images of satirical skeletons still common in Mexico today. Around the Day of the Dead, Posada used comical images of the upper-class and politicians as skeletons to portray death as a great equalizer of injustice.

The skull was an important symbol of death and sacrifice in the pre-Colombian period, but the figure of a satirical and comic death is a more recent phenomenon. La Catrina, a female dandy, was one of many images created by Posada. La Catrina dolls have become an icon of The Day of the Dead.

Depictions of mummies are everywhere – trinkets, cartoons, candies, art galleries, fine jewelry, literature, street festivals and in songs. During the early 1970s, the mummies battled masked Mexican wrestlers in films, including *"The Santo Versus the Mummies of Guanajuato."* Another wrestling / mummies film added a mad scientist and killer dwarfs.

The way the Detroit Science Center chooses to honor the mummies is by telling their stories as thoroughly as possible. To honor the beautiful city where they reside and the remarkable culture that exists in Guanajuato. And to advance the wonderful science opportunities inherent with the mummies and share that information with the masses. The Accidental Mummies project would not be possible without Detroit Science Center President & CEO Kevin Prihod, who early on understood the educational and cultural value of the mummies.

In Mexico, El Museo de las Momias Director Juan Manuel Guerrero and Dr. Eduardo Romero Hicks, then mayor of the city, played crucial roles in allowing the 36 mummies to travel to the U.S. for the first time.

EL Museo de las Momias Director Juan Manuel Guerrero

The exhibit's guiding force has been Martina Guzman, a native Detroiter and proud Chicana. She ensured the cultural authenticity of the exhibit and applied sheer force of will to finalize the deal between the city of Guanajuato and the Science Center. She played an essential role in the design and layout of the exhibit. She spent months collecting artifacts and hours fine-tuning a plethora of details inside the exhibit. She led a film crew that will produce a documentary of Guanajuato and the mummies. And she provided support for the creation of this book.

The book would not look as beautiful as it does without Michael Ulinski and Christina Pattyn of Ciel Design Partners and Kelly Fulford, vice president of sales and marketing for the Detroit Science Center. Vivian Henoch, the Detroit museum's medical science content developer, was the major force in explaining the science in the exhibit. She acted as the main liaison to mummy experts Ronald Beckett and Jerry Conlogue.

Oakwood Imaging Center of Dearborn, Mich., generously allowed the use of their CT Scans. Siemens Healthcare provided the amazing images of the mummies. The depth of detail in the exhibit has been produced by Detroit Science Center Design & Exhibits at Eekstein's Workshop in Ferndale, Mich. The Detroit Science Center is honored to present *The Accidental Mummies of Guanajuato*.

Long Live the Dead.

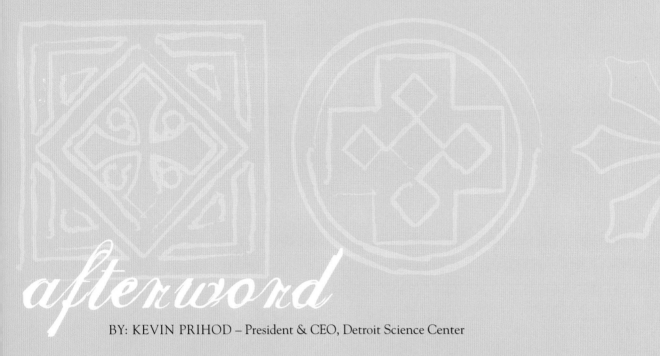

afterword

BY: KEVIN PRIHOD – President & CEO, Detroit Science Center

This book heralds an important museum exhibit that is the result of a collaboration of two great museums from two great cities in two great countries.

The Accidental Mummies of Guanajuato features 36 accidental mummies on loan from El Museo de las Momias in Guanajuato, Mexico, that have never been seen outside their native country. It combines science, history and cultural anthropology and immerses visitors in the world of a Mexican city over 100 years ago where deceased residents naturally mummified in their crypts.

Only 1 in 100 bodies entombed in Guanajuato experienced this rare and mysterious process of natural mummification. Visitors to the exhibit will meet these accidental mummies, learn about life in their thriving community, discover the modern-day forensic technology that helps scientists analyze them and explore a culture that reveres and celebrates them.

The Accidental Mummies of Guanajuato is a once-in-a-lifetime opportunity to see these amazing mummies in the United States. The limited three-year tour begins at the Detroit Science Center on October 10, 2009.

At the Detroit Science Center, we are truly honored the city of Guanajuato chose us to design and build this important exhibit and protect their treasured mummies as they tour the United States.

To the people of Guanajuato, the mummies are very special. They are loved as family members and friends, honored for their importance to the community and respected as ambassadors of goodwill. The decision to allow them to leave their city to tour the United States was a difficult one for the elected officials.

Dr. Eduardo Romero Hicks was mayor of Guanajuato from 2006 to 2009. He has supported and directly participated in the remodeling of the Museo de las Momias and the forensic study of the mummies, along with teams of experts from the United States. He is a general surgeon, trained in Mexico and the United States.

We are honored to have three of the most renowned scientists specializing in mummy forensics assisting us on this project.

Dr. Ronald Beckett is a bioanthropologist from Quinnipiac University.

Gerald Conlogue is a radiologist from Quinnipiac University and both gentlemen served as hosts of the popular National Geographic television series, *The Mummy Road Show*.

Dr. Jerry Melbye is a distinguished forensic anthropologist from University of North Texas who has studied the mummies of Guanajuato extensively.

Oakwood Healthcare System is one of Michigan's premier healthcare providers. They generously welcomed our mummies into their Imaging Center in Dearborn, Mich., and conducted full-body CT scans of seven of the mummies that have produced spectacular images and provided information to help us understand how these mummies lived and died.

Siemens Healthcare is one of the world's largest providers of medical technology. They have assisted us with the processing and interpretation of the CT images. In addition, they have provided a state-of-the-art 64 slice dual-source CT Scanner that will serve as the centerpiece of our exhibit's Forensics Lab that will accompany the exhibit on its national tour.

Barbara Martin Bailey is one of the nation's most prominent FBI-certified forensic artists. Her work has helped solve murders, identify missing persons and capture criminals. She has produced drawings and clay models of our mummies so visitors to the exhibit will see these mummies as they looked and lived 100 years ago in Guanajuato.

The Accidental Mummies of Guanajuato is an extraordinary exhibit and we invite you to experience it.

credits

ILLUSTRATION:
© Barbara Martin Bailey / Christine Chambers: 117, 123, 125, 127, 129, 131, 133, 135, 137, 139

PHOTOGRAPHY:
© Edward Gorczyk: 2–3, 10 (INSETS 1+3), 11, 12, 16–17, 18–19, 22–23, 24 (INSET), 26–27, 30, 31, 32–33, 36–37, 52–53, 56–57, 58, 59, 60, 61, 64–65, 71, 74–75, 80–81, 82, 84–85, 87–105, 114, 122, 124, 126, 128, 130, 132, 134, 136, 138, 142–143, 144, 145, 146, 147, 148, 154, 155

© Janet Jarmon: COVER, 6–7, 8–9, 10 (INSET 2), 20–21, 24–25, 28–29, 34, 35, 38–39, 40, 41, 42–43, 44–45, 46, 47, 48, 49, 50–51, 55, 62–63, 66–67, 72–73, 76–77, 78–79, 140–141

© Vivian Henoch: 14–15, 110–111, 112, 115, 120, 121

© Julio Zenil: 68–69

CT Image Processing:
Oakwood Healthcare System / Siemens Healthcare: 106, 107, 108, 116, 118, 119

AUTHOR'S ACKNOWLEDGMENTS:
Louis Aguilar owes a huge amount of gratitude to Martina Guzman, the Detroit Science Center staff, Ciel Design Partners and the very cool city of Guanajuato. He needs to also thank Marisela Riveros, Courtney Smith, Frances Sanchez, Arlene Aguilar and all his friends as well his colleagues at *The Detroit News* who offered much emotional support.

The Accidental Mummies of Guanajuato is produced by Detroit Science Center Design & Exhibits at Eekstein's Workshop, LLC, a wholly owned subsidiary of the Detroit Science Center, in association with Accidental Mummies Touring Company LLC.

The Accidental Mummies of Guanajuato Logo/Trademark™ is Copyright © Accidental Mummies Touring Company LLC.

The producers of this book have made every effort to ensure the accuracy of the information contained herein. If any errors unwittingly occurred, we will be happy to correct them in future editions.

FORENSIC ANTHROPOLOGY TEAM

Quinnipiac University
Ronald G. Beckett
Professor Emeritus
Department of Biomedical Sciences
Executive Co-Director of the
Bioanthropology Research Institute

Gerald Conlogue
Professor of Diagnostic Imaging
Executive Co-Director of the
Bioanthropology Research Institute

Jiazi Li
Research Assistant

Oakwood Healthcare System
Lisa Bain
Gregory Bock
Emad Hamid, M.D.
Judith McNeeley
Henry Pierson
Timothy Vargas
Mary Zatina

Siemens Healthcare
Roselle Charlier
James Forman

Center for Human Identification Laboratory of Forensic Anthropology University of North Texas
Dr. Jerry Melbye, D-ABFA, F-AAFS
Research Professor

Vicky Melbye
Research Assistant

EXHIBITION TEAM

American Exhibitions, Inc.
Boca Raton, Florida

Detroit Science Center
Kerri Budde
Jennifer Clark
Joshua Dawson
Kelly Fulford
Velda Garcia
Edward Gorczyk
Martina Guzman
Pete Herb
Amanda Jackson
Julie Johnson
Tom Mott
Kevin Prihod
Paul Rossen
Todd Slisher

Detroit Science Center Design & Exhibits
Eli Baderca
Gav Baderca
Brian Balagna
Joe Balagna
Andrew Ball
Pat Barrie
Don Blaker
Don Bogart
Justin Boudrie
Elizabeth Chilton
Mike Cronyn
Theresa DeRoo
Shawn Dlabal
Sarah Doan
Albert Barney Einstein
Jeff Evarts
John Everett
Vic Fischione

Vivian Henoch
Julie Hinzmann
Lisa Hughes
Andrew Hursin
Robert Hursin
Gary Jennings
Doug Johnson
Mike Kendra
Pat Mackler
Dik Moore
Susan Morgowicz
Pat Nowland
Taressa Owens
Steve Reid
Dave Ricca
Gary Rizzo
Laurel Robinson
Robert Seestadt
Ray Slowik
Al Spreeman
Ed Summers
Sarah Tanner
Josh Viola
Pat Weiss
Rob Wethington
James Wilder

Marcon Exhibits & Events
Robert L. Gardella

ADDITIONAL SUPPORT PROVIDED BY:
Erickson Labs Northwest
George Salem
Karl Storz
Michigan State Police
   Forensic Lab
PKA Design
Ron Novak
VCA Veterinary Clinics

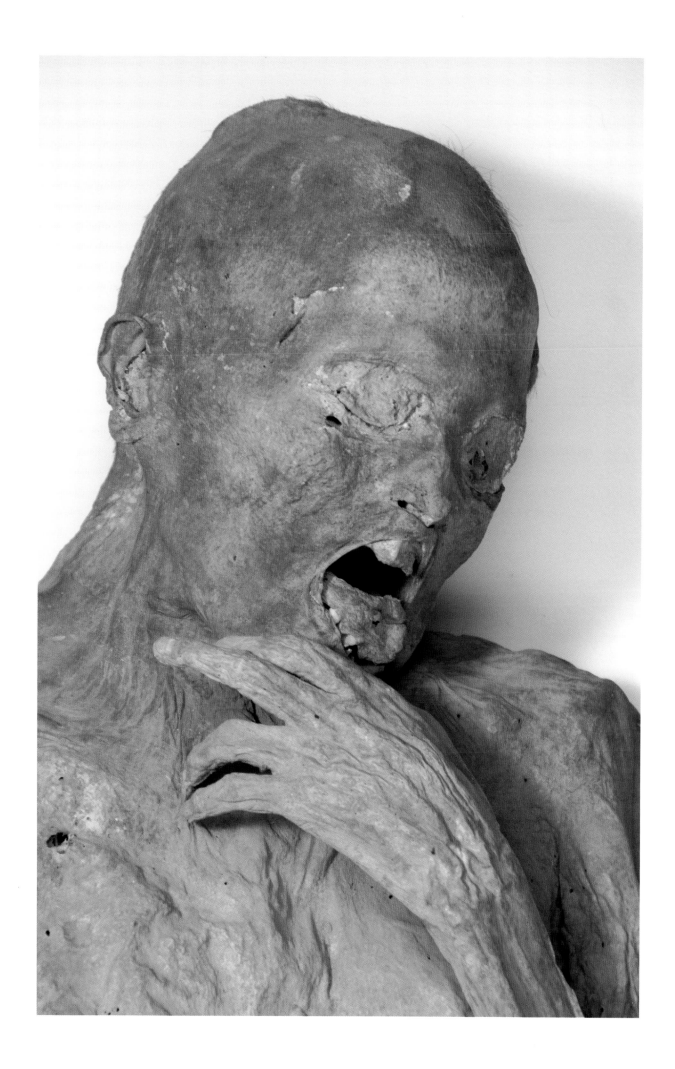